Metaheuristics for Enterprise Data Intelligence

With the emergence of the data economy, information has become integral to business excellence. Every enterprise, irrespective of its domain of interest, carries and processes a lot of data in their day-to-day activities. Converting massive datasets into insightful information plays an important role in developing better business solutions. Data intelligence and its analysis pose several challenges in data representation, building knowledge systems, issue resolution and predictive systems for trend analysis and decision-making. The data available could be of any modality, especially when data is associated with healthcare, biomedical, finance, retail, cybersecurity, networking, supply chain management, manufacturing, etc. The optimization of such systems is therefore crucial to leveraging the best outcomes and conclusions. To this end, AI-based nature-inspired optimization methods or approximation-based optimization methods are becoming very powerful. Notable metaheuristics include genetic algorithms, differential evolution, ant colony optimization, particle swarm optimization, artificial bee colony, grey wolf optimizer, political optimizer, cohort intelligence and league championship algorithm. This book provides a systematic discussion of AI-based metaheuristics application in a wide range of areas, including big data intelligence and predictive analytics, enterprise analytics, graph optimization algorithms, machine learning and ensemble learning, computer vision enterprise practices and data benchmarking.

Dr Kaustubh Vaman Sakhare, Sr. Data Scientist, System Engineering & Production Integration (SEPI), John Deer, Pune, India.

Dr Vibha Vyas, Associate Professor, Department of Electronics and Telecommunication, College of Engineering, Pune, India.

Dr Apoorva S Shastri, Research Assistant Professor, Institute of Artificial Intelligence, MIT World Peace University, Pune, India.

Advances in Metaheuristics

Series Editors:
Patrick Siarry, Universite Paris-Est Creteil, France
Anand J. Kulkarni, Institute of Artificial Intelligence, MIT World Peace University (MIT-WPU), India

Metaheuristics for Enterprise Data Intelligence
Edited by Dr Kaustubh Vaman Sakhare, Dr Vibha Vyas, Dr Apoorva S Shastri

Handbook of AI-based Metaheuristics
Edited by Patrick Siarry, Anand J. Kulkarni

Metaheuristic Algorithms in Industry 4.0
Edited by Pritesh Shah, Ravi Sekhar, Anand J. Kulkarni, Patrick Siarry

Constraint Handling in Cohort Intelligence Algorithm
Ishaan R. Kale, Anand J. Kulkarni

Hybrid Genetic Optimization for IC Chip Thermal Control: With MATLAB® applications
Mathew V K, Tapano Kumar Hotta

Handbook of Moth-Flame Optimization Algorithm: Variants, Hybrids, Improvements, and Applications
Edited by Seyedali Mirjalili

Combinatorial Optimization Under Uncertainty: Real-life Scenarios in Allocation Problems
Edited by Ritu Arora, Prof. Shalini Arora, Anand J. Kulkarni, Patrick Siarry

AI-based Metaheuristics for Information Security in Digital Media
Edited by Apoorva S Shastri, Mangal Singh, Anand J. Kulkarni, Patrick Siarry

For more information about this series please visit: https://www.routledge.com/Advances-in-Metaheuristics/book-series/AIM

Metaheuristics for Enterprise Data Intelligence

Edited by
Kaustubh Vaman Sakhare, Vibha Vyas
and Apoorva S Shastri

CRC Press
Taylor & Francis Group
Boca Raton London New York

CRC Press is an imprint of the
Taylor & Francis Group, an **Informa** business

Designed cover image: Shutterstock Images

First edition published 2025
by CRC Press
2385 NW Executive Center Drive, Suite 320, Boca Raton FL 33431

and by CRC Press
4 Park Square, Milton Park, Abingdon, Oxon, OX14 4RN

CRC Press is an imprint of Taylor & Francis Group, LLC

ISBN: 978-1-032-68377-5 (hbk)
ISBN: 978-1-032-69979-0 (pbk)
ISBN: 978-1-032-69980-6 (ebk)

DOI: 10.1201/9781032699806

Typeset in Minion Pro
by Apex CoVantage, LLC

Contents

Preface

WITH THE EMERGENCE OF the data economy, information has become the key to business excellence. Every enterprise, irrespective of its domain of interest, carries and processes a lot of data in their day-to-day activities. Converting massive datasets into insightful information plays an important role in developing better business solutions. Data intelligence and its analysis pose several challenges in data representation, building knowledge systems, issue resolution and predictive systems for trend analysis and decision-making. Optimization of such systems is very much crucial to leverage the best outcome and conclusion. As the data available could be of any modality especially when data is associated with healthcare, biomedical, finance, retail, cybersecurity, networking, supply chain management, manufacturing and so on. AI-based nature-inspired optimization methods or approximation-based optimization methods are becoming very popular to gain optimum solutions in the business. The notable metaheuristics are genetic algorithms, differential evolution, ant colony optimization, particle swarm optimization, artificial bee colony, grey wolf optimizer, political optimizer, cohort intelligence and league championship algorithm, just to mention a few. This book intends to provide a systematic discussion on the application of AI-based metaheuristics in the following areas and more. This book can be a baseline reference to the scientist, professors and research scholars working in the domain of enterprise data intelligence. The individual contributions appearing in this book are the following:

- **Chapter 1**, "Terror Attacks Forecast Using Machine Learning and Neo4j Sandbox: A Review", highlights the predictability of terror attacks using machine learning algorithms. The review unravels the fact that terror attacks can indeed be perceived as one of the most challenging issues and an influential factor in the social and economic development of any nation. Before the surge of machine learning, it was obvious to consider terror attacks as random phenomenon. But after diving deep into the datasets like the Global Terrorism Database, it was deciphered that there is an intricate pattern involved in the occurrence of the terror attacks. The review gives out a proper idea on the use of machine learning techniques to predict future occurrences of such attacks, to save many innocent lives.
- **Chapter 2**, "5G Evolution and Revolution: A Study", presents a survey on evolution of 5G and the revolution in the realm of technology. The study incorporates various

key characteristics which will surely transform the future mobile communication technology for ever. The evolution of communications over various generations has been discussed to bring the systematic emergence of telecommunication technology. Various protocols are presented to get a closer understanding of 5G communication. The chapter presents different advantages and applications to summarize the discussion.

- **Chapter 3**, "Metaheuristic Algorithms and Its Application in Enterprise Data", similarly titled to this book, entails various heuristics algorithms for enterprise data while also throwing light on topics like role and importance of enterprise data, architecture of AI-based model for enterprise data and how it is deliberated with necessary magnitude. The need of metaheuristics in enterprise data analysis and various algorithms and their prominent use in enterprise data handling is majorly visualized in this chapter.
- **Chapter 4**, "Petrographic Image Classification Accuracy Improvement Using Improved Learning", encompasses the concept of petrographic image classification based on the microstructural imaging of the mineral rock by using the improved machine learning algorithms. The author uses geologist-classified grain-size mineral rocks' microstructure images as the input data for algorithms. The algorithms used for the experimentation and analysis under this chapter are optimized version of radial basis function (RBF) SVM, Adaboost algorithm and convolutional neural networks classifiers. The chapter also introduces an iterative improved learning novel approach for the improvement of model accuracy by enhancing the learning knowledge acquired by the model during the training process.
- **Chapter 5**, "Data Visualization and Dashboard Design for Enterprise Intelligence", presents the importance of information representation in visual format. The use of various charts and graphs makes the multi-fold interpretation of the data. The use of augmented reality and machine learning in the enterprise data dashboard design has become a need of organizational competency. The use of various techniques and tools in the design and their extended use in different applications are also discussed.
- **Chapter 6**, "Beyond the Hype: Understanding the Potential of ChatGPT in the Articulation of Technical Papers", evaluates the use of ChatGPT in writing technical papers. Compared to other applications, the use of ChatGPT in technical paper writing needs technical accuracy and creative expression. The lack of originality, ethical considerations and the need of human supervision are some of the trivial issues, impacting the performance of the solutions deeply. The chapter compares various AI-powered language generation models. The study also encourages prospective research in this domain by quoting acceptance of ChatGPT by various publication houses.
- **Chapter 7**, "Metaheuristics and Deep Learning in Lung Nodule Detection and Classification", brings a comprehensive survey on lung cancer and the various

biological abnormalities causing this life-taking disease. A systematic computer vision approach to understand the various stages of lung nodules is discussed with the help of respective images. Beginning from computer vision-based lung nodule detection up to deep learning based and metaheuristic approaches for deep learning models are the purview of the deep research presented in this chapter.

- **Chapter 8**, "An Improved Face Recognition Method Using Canonical Correlation Analysis", presents thorough in-depth research in the domain of facial recognition, which is the most used recognition method in the world. As known, face recognition is useful in various biometric applications and security systems. In this chapter a detailed discussion is made to illustrate the process of face recognition with the help of canonical correlation analysis and local binary pattern method.
- **Chapter 9**, "Guesswork to Results: How ML-Based A/B Testing Is Changing the Game", entails the machine-learning-based technique to determine the effectiveness of digital marketing initiatives. The chapter brings certain interesting aspects of the use of machine learning in the well-known A/B testing and compares it thoroughly with traditional statistical methods and showcases how optimizing one's digital marketing can directly influence the company's sales and growth. The discussion illustrates the proposition with the help of a case study.

Contributors

Krishnansh Awasthi
Bharati Vidyapeeth (Deemed to be a University), College of Engineering Pune
Pune, Maharashtra, India

Bhushan M. Borhade
Department of AI & ML Engineering, Navsahyadri College of Engineering, Savitribai Phule Pune University
Pune, Maharashtra, India

Bhavay Chopra
Bharati Vidyapeeth (Deemed to be a University), College of Engineering Pune
Pune, Maharashtra, India

Aditya Choudhary
Bharati Vidyapeeth (Deemed to be a University), College of Engineering Pune
Pune, Maharashtra, India

Rushikesh Dudhani
COEP Technological University
Pune, Maharashtra, India

Anwar Hussain
Critical River Technologies Pvt Ltd, Hyderabad, Telangana, India

Ganesh D. Jadhav
Department of Computer Engineering, SKNCOE Vadgaon, Savitribai Phule Pune University
Pune, Maharashtra, India

Radhika D. Joshi
COEP Technological University
Pune, Maharashtra, India

Payal Kadam
Bharati Vidyapeeth (Deemed to be a University) College of Engineering
Pune, Maharashtra, India

Rama Vaibhav Kaulgud
Sanjay Ghodawat University
Kolhapur, Maharashtra, India

Suchitra Khoje
CS2S Technology Pvt. Ltd.
Pune, Maharashtra, India

Ashutosh Marathe
Vishwakarma Institute of Technology
Pune, Maharashtra, India

Chetan More
Bharati Vidyapeeth (Deemed to be a University) College of Engineering
Pune, Maharashtra, India

Suhas Patil
Department of Computer Engineering, Bharti Vidyapeeth University, College of Engineering
Pune, Maharashtra, India

Vinod Patil
Bharati Vidyapeeth (Deemed to be a University) College of Engineering
Pune, Maharashtra, India

Ankit Raj
MIT WPU, Kothrud
Pune, Maharashtra, India

Mandar Saundattikar
San Jose State University
Fremont, California, USA

Neha Shaah
Central Analytics, Tata Motors Ltd.
Pune, Maharashtra, India

Namita K. Shinde
Bharati Vidyapeeth (Deemed to be a University) College of Engineering
Pune, Maharashtra, India

Sagar Shinde
Dr. D. Y. Patil Institute of Technology
Pimpri, Pune, Maharashtra, India

Yogesh Shinde
Department of Computer Engineering, AVCOE Sangamner, Savitribai Phule Pune University
Pune, Maharashtra, India

Bahubali Shiragapur
D. Y. Patil International University,
Akurdi, Pune, Maharashtra, India

Nishikant Bhaskar Surwade
Critical River Technologies Pvt Ltd,
Hyderabad, Telangana, India

Tanuja Tewari
College of Engineering
Pune, Maharashtra, India

Falguni Vyas
New York University
Department of Electrical Engineering, San Jose State University,
Fremont, California, United States

Lalitkumar Wadhwa
Dr. D. Y. Patil Institute of Technology
Pimpri, Pune, Maharashtra, India

Sheetal Waghchaware
COEP Technological University
Pune, Maharashtra, India

CHAPTER 1

Terror Attacks Forecast Using Machine Learning and Neo4j Sandbox

A Review

Sagar Shinde, Suchitra Khoje, Ankit Raj and Lalitkumar Wadhwa

1.1 INTRODUCTION

The current scenario in which the whole world's peace and economic growth is endangered due to the rise of the Taliban in Afghanistan is a result of ignorance of world powers about terrorism. Countries which are not affected by terrorism are not easy to find in today's era. The USA, the European Union, India, China, Russia and even the African continent are under the influence of terrorist organizations like Boko Haram, AQ, Taliban, ISIS – K and many more. The attacks are well planned and do have a pattern of occurrence. The Global Terrorism Database (GTD) shows us that from the 1970s to the current time there have been around 200,000 terror attacks. This shows terrorism is still one of the biggest hurdles in the path of peace and prosperity. In this chapter, I will be reviewing the work done to date to stop this catastrophic event in the domain of machine learning and artificial intelligence. I will also be including my viewpoints regarding the topic and what else can be done to make this technology even more firm and efficient.

1.1.1 Machine Learning

Machine learning is a technique of making computer algorithms intelligent by training them on datasets. We can say that machine learning is the initial and primitive stage of artificial intelligence, but it is highly dependent on data engineering, which means that the more the data would be clean the better results we would get from machine learning models. It is widely used in data science applications, but nowadays it is involves multiple domains. The influence of machine learning spans from computer science to the automotive domain and from manufacturing to the space domain (Figure 1.1).

DOI: 10.1201/9781032699806-1

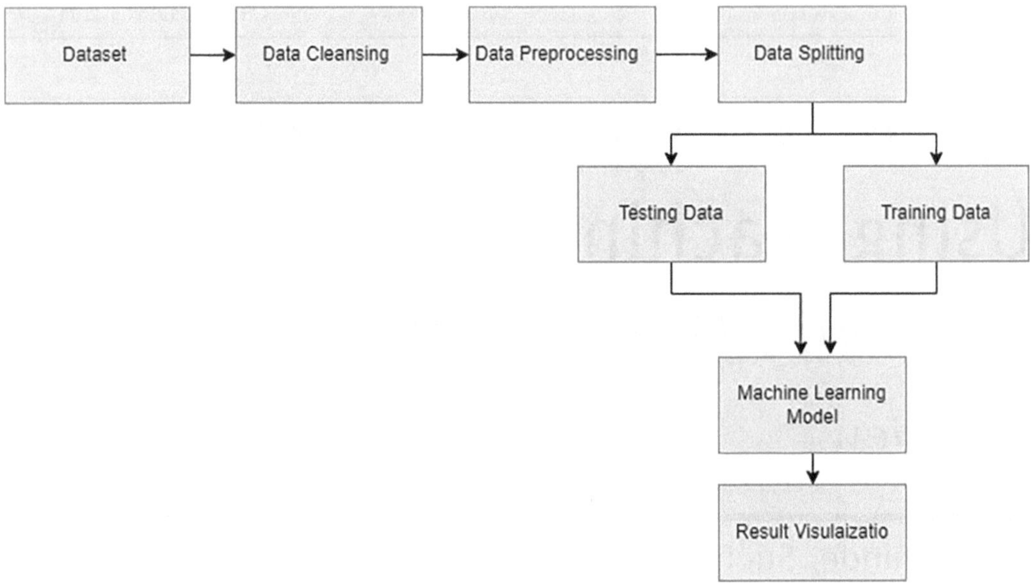

FIGURE 1.1 Block diagram of a generic machine learning model.

1.1.2 Deep Learning

We can say that deep learning is an advanced version of machine learning in deep learning we don't have a separate step for feature extraction and classification, unlike machine learning. On the robustness parameter, we can say that deep learning is more robust than machine learning techniques. As we know, with higher performance comes higher resource allocation in combination; hence, deep learning algorithms will require more processing power than machine learning algorithms (Figure 1.2).

1.1.3 Global Terrorism Dataset Description

The GTD is a database of all the terror attacks and attempts of terror attacks from 1970 to 2020 in a combined form. It gives insights into the attack type, attack weapon type, location, time, year, causality and many more, which help us understand the psyche of terrorists and anarchists and their attack patterns. It is also updated from time to time and is maintained by START, which is an organization dealing with such databases. The GTD has around 200,000 entries and approximately 130 attributes. It consists of the day, month and year for all the incidents; it includes the country and region of the incident. It also has

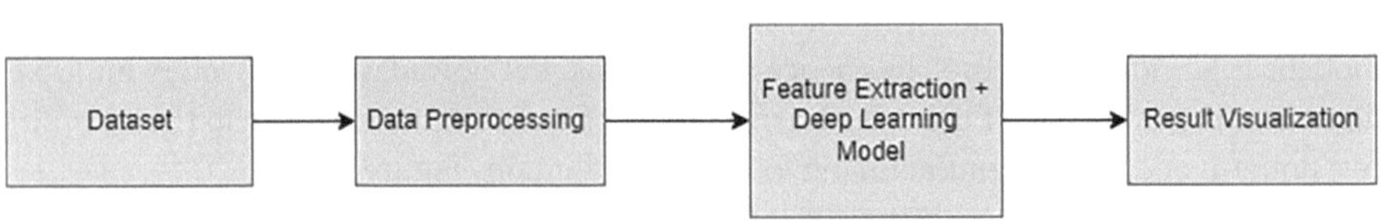

FIGURE 1.2 Block diagram of a generic deep learning model.

the attack type, including assassination, kidnapping, bombing, armed assault, hijacking, etc. The database has the target type attribute, which includes private citizen, government diplomat, journalist and media, police, utilities, military, etc. The GTD also has weapon-type attribute, including entries like firearm, incendiary and explosive.

1.2 LITERATURE SURVEY

The researcher was able to get an accuracy of about 90% for multiple machine learning models, including the decision tree, bagging classifier, random forest, some extra tree classifiers and XGBoost classifiers (Table 1.1) [1]. The precision and recall rate were also above or equal to 90%, but there was an observation made that as the number of terror attacks decreased in range, the accuracy fell from 90% to 79%. The precision also fell from 97% to 12% for classifiers like XGBoost. This shows that the number of entries in the GTD affects the accuracy output of the machine learning models [1]. Here the researchers applied random forest on the GTD in the form of classification as well as regression for different attributes. For attack, type attributes the accuracy and regression are maximum whereas for the target type attributes the accuracy and regression are minima so we can say that the attack type attributes in the GTD are more dominant than any other attributes present in the database [2]. In Ref [2], the researcher has applied an ensemble type of machine learning algorithm using the feature selection techniques, such as chi-square feature selection,

TABLE 1.1 Literature Survey

Serial No.	Publishing Year	Platform Used	Features Extracted	Database	Models
1	2021	Python	36	GTD	Decision Tree Bagging Classifier Random Forest Extra Trees XGBoost
2	2017	Python	4	GTD	Random Forest
3	2021	Python		GTD	SVM & KNN Ensemble Model
4	2020	Python	46	GTD	Decision Tree Random Forest
5	2020	Python, Twitter 4j	Not Specified	Twitter trending hashtag capturing	KNN SVM
6	2020	Python	48	GTD	KNN
7	2020	Python	32	GTD	XGBOOST
8	2020	Python	8	GTD	Naïve Byes MLP Decision Tree
9	2020	Python	34	GTD	NN DNN
10	2020	Python, Mexican Database	33	Mexican warehouse database	SOM Neural Gas C-Means

information gain feature selection and hybridized feature selection, for which they were able to achieve an accuracy of 94.17%, 97.34% and 97.81%, respectively. An ensemble type of machine learning model is a combination of two or more two machine learning models to increase the efficiency of the output [3]. Here the researcher has tried to implement decision tree as well as random forest machine learning models for global terror attack prediction. For decision trees, they were able to achieve an accuracy of 75% approximately, and with further modifications, they were able to increase the accuracy to 79.24%, while for the random forest, they were able to achieve an accuracy of 89.54% [4]. The data collection technique in [4] is a bit different because the source of data here is the Twitter trending hashtags in the related context. The tool used for data collection was Twitter 4j application peripheral interface. The machine learning models applied here are the k-nearest neighbour algorithm and support vector machine algorithm for which they were able to achieve an accuracy of 50% and 44.39%, respectively. They applied the machine learning model on another set of data but no noticeable change in the accuracy of both machine learning models was observed. Here the researchers tried to implement three different types of machine learning models for global terror attack prediction using the GTD. The models implemented by them are random forest, decision tree and KNN, but the result achieved by applying these models were not clearly defined and stated by the author [5]. The XGBoost machine learning model was implemented by the author in [5], and with the help of a combined decision tree and gradient boosting, he was able to achieve an accuracy of 86.5%. The same dataset was applied four different machine learning models, such as decision tree, random forest, logistic regression and naïve Bayes, but it was observed that the accuracy achieved from the XGBoost model surpassed the accuracy achieved by other models not only that the sensitivity, the precision rate and even the F1 score of the XGBoost model what's surpassing other machine learning models [6]. The researcher here has tried to implement multiple machine learning classifier models including naive Bayes, multi-layer perceptron, multi-class classifier, lazy classifier, decision tree, etc. The accuracy for each classifier ranged from 90% to 93.44%. With further modifications, the accuracy could have been reached to around 99% [7]. With the help of the GTD, the author tries to implement neural network and deep neural network models. The accuracy for the neural network and deep neural network was around 86% and 92%, respectively. The accuracy for both the models remains unchanged for around 200 iterations and there were no significant changes in inaccuracy for around 100 iterations, but it was observed that after 500 iterations accuracy of both the models was almost equivalent [8]. As the database for this research, what's the real-time database from a Mexican data warehouse the accuracy for the applied model which is an ensemble machine learning model, which included C-means clustering, strict competitive learning, self-organizing map and the neural gas network was 66.66%. The accuracy might have been affected by the presence of real-time data [9, 10]. Numerous applications of the graph database exist, including network administration, social connectedness, biology and the detection of dishonest behaviour. It provides the developer with greater performance, flexibility and agility compared to relational databases [11]. With results ranging from 87% to 97%, hybrid models and ensemble machine learning algorithms are also able to achieve promising results. The performance metrics employed in the result analysis, which is mostly based on ROC curve analysis for each model, include ROC, AUC, precision, recall and F1 score. The prediction accuracy of machine learning algorithms is improved with the use of metaheuristic optimization techniques [12].

1.3 MODELS

A. **Decision Tree:** A decision tree is a machine learning classifier model which classifies datasets based on entropy and information gain. The major advantage of using a decision tree algorithm is that it does not require any data normalization technique or even data scaling technique. Even though a decision tree can handle categorical data it is highly unstable since a small change in the input can bring a drastic change in the output. The training time required for the high amount of data is also more compared to other machine learning algorithms [1–3].

B. **Bagging Classifier:** A bagging classifier is a collective, or we can say ensemble type of machine learning algorithm, in which models of similar algorithms are trained together on the common set of data. It is generally used with a decision tree to make it more effective and accurate. The bagging classifier reduces the chance of overfitting occurring but one of its major drawbacks is its complexity and high computational power requirement [2, 3].

C. **Random Forest:** Random forest is a bagging type of classifier, which is a combination of different decision tree algorithms trained on a common random dataset. It reduces the variance in the dataset making the overfitting problem disappear. Although random forest is a very good classifier, its performance can be increased using the hyperparameter tuning method [1–4, 13].

D. **XGBoost:** XGBoost is a combination of bagging classifier tree algorithm and gradient boosting algorithm which enhances the accuracy of the decision tree algorithm. The flexibility of this algorithm is more, but it can handle non-linear and unstructured data very well [1, 8].

E. **KNN:** KNN, or k-nearest neighbour, is one of the most effective classification algorithms used for data mining applications. KNN can be bought into both applications, which is into the regression application and in the classification algorithm. Despite its advantages, it is a bit slower in the terms of learning [7, 8].

F. **SVM:** SVM, or support vector machine, is a supervised learning algorithm that classifies the data based on the hyperplane. It is very good for performing on low-memory devices but the SVM algorithm does not work for extensive datasets [7, 8].

G. **Multilayer Perceptron:** Multilayer perceptron is a neural network that works for both bipolar and unipolar inputs and gives output according to the activation function implemented in the model. Multilayer perceptron can handle a large amount of data but its major limitation is that it can't process categorical data. It can work for both linear and non-linear types of data [10].

H. **Logistic Regression:** Logistic regression is a probabilistic approach to predict an event in the form of yes or no, true or false and even one or zero. The input of logistic regression has no range limitation that is, it can range from infinity to the counterpart of infinity, but the output can lie between zero and one only [7, 8].

I. **Neural Network:** A neural network is a group of algorithms which does perform data classification as well as data clustering. The major advantage of a neural network is that it is very fault-tolerant and has concurrent processing ability, but it may require high computational power and hardware resources to do so [6, 8, 13].

J. **Deep Neural Network:** Deep neural network is an extension of a neural network that can process higher-dimensional data such as images and videos. It has an adequate amount of flexibility which is difficult to find in neural networks, but it is not that cost-efficient, which means it can be expensive at times [6, 8, 13].

1.4 REMARKS

The work done to date for global terror attack prediction using machine learning is based on the GTD, which is a relational database. With the emergence of new technologies like graph databases and deep learning, the prediction for terror activities could be made more efficient and robust. We can create the database using platforms like Neo4j sandbox and tiger graph and could extract features like graph embedding and node embedding and can use these features as an input two various machine learning models and even deep learning models. Although the graph database creation of a massive database like the GTD would be difficult dusk due to the presence of various null values and redundant features in the database. We could also refer to other databases if available to check the efficiency and accuracy of different machine learning models on different datasets.

1.5 CONCLUSIONS

The Global Terrorism Database is a very massive database, but the fact that can't be ignored is that it has a very high number of redundant attributes that add null values. To handle such an inefficient database, software like Neo4j should be used so that with the help of its huge graph data science library, the null values from the database would be removed before feeding this data to a machine learning model. This would make the prediction model much more efficient and could also increase the recall rate and precision of the model. The considerable features like graph embedding, page rank, degree and centrality from the graph and apply these extracted features on deep learning algorithms as well. The most promising methods for classifying terror events using the orange data mining tool are random forest and gradient boosting. Graph embedding improved the model's predictive performance and made up for the inconsistent dataset.

REFERENCES

1. Pan, Xiaohui. (2021). "Quantitative Analysis and Prediction of Global Terrorist Attacks Based on Machine Learning", *Scientific Programming*, 2021, 15. Article Id 7890923.
2. Saha, Snehanshu, Aladi, Harsha and Kurian, Abu. (2017). "Future Terrorist Attack Prediction using Machine Learning Techniques". https://doi.org/10.13140/RG.2.2.17157.96488

3. Olabanjo, Olusola A., Aribisala, Benjamin S., Mazzara, Manuel and Wusu, Ashiribo S. (2021). "An Ensemble Machine Learning Model for the Prediction of Danger Zones: Towards a Global Counter-Terrorism", *Soft Computing Letters*, 3, 100020. https://doi.org/10.1016/j.socl.2021.100020; ISSN 2666-2221
4. Huamaní, Enrique Lee, Alicia, Alva Mantari and Roman-Gonzalez, Avid. (2020). "Machine Learning Techniques to Visualize and Predict Terrorist Attacks Worldwide using the Global Terrorism Database", *International Journal of Advanced Computer Science and Applications (IJACSA)*, 11(4). https://doi.org/10.14569/IJACSA.2020.0110474
5. Narula, Meenakshi, Kumar, Siddharth, Kaul, Saksham and Ks, Saurav. (April 6, 2020). "Predictive Analysis on Global Terrorism", Proceedings of the International Conference on Innovative Computing & Communications (ICICC) 2020. https://doi.org/10.2139/ssrn.3569598; https://ssrn.com/abstract=3569598
6. Feng, Y. et al. (2020). "An XGBoost-Based Casualty Prediction Method for Terrorist Attacks", *Complex & Intelligent Systems*, 6, 721–740. https://doi.org/10.1007/s40747-020-00173-0
7. Kumar, Vivek, Mazzara, Manuel, Messina, Angelo. (2018). "A Conjoint Application of Data Mining Techniques for Analysis of Global Terrorist At-tacks Prevention and Prediction for Combating Terrorism", 6th International Conference in Software Engineering for Defense Applications-SEDA 2018, Rome, Italy. https://doi.org/10.48550/arXiv.1901.06483
8. Uddin, M. Irfan et al. (2020). "Prediction of Future Terrorist Activities Using Deep Neural Networks", *Complexity*, 2020. Article Id 1373087.
9. Rocha-Salazar, José-de-Jesús, Segovia-Vargas, María-Jesús and Camacho-Miñano, María-del-Mar. (2021). "Money Laundering and Terrorism Financing Detection Using Neural Networks and an Abnormality Indicator", *Expert Systems with Applications*, 169, 114470. https://doi.org/10.1016/j.eswa.2020.114470; ISSN 0957-4174
10. Macák, Martin, Stovcik, Matus, Buhnova, B. (2020). "The Suitability of Graph Databases for Big Data Analysis: A Benchmark", 5th International Conference on Internet of Things, Big Data and Security. https://doi.org/10.5220/0009350902130220
11. Timón-Reina, S., Rincón, M. and Martínez-Tomás, R. (May 18, 2021). "An Overview of Graph Databases and Their Applications in the Biomedical Domain", *Database (Oxford)*, 2021, baab026. https://doi.org/10.1093/database/baab026; PMID: 34003247; PMCID: PMC8130509
12. Python, A., Bender, A., Nandi, A.K., Hancock, P.A., Arambepola, R., Brandsch, J. and Lucas, T.C.D. (July 30, 2021). "Predicting Non-State Terrorism Worldwide", *Science Advances*, 7(31), eabg4778. https://doi.org/10.1126/sciadv.abg4778; PMID: 34330703; PMCID: PMC8324061
13. Sarker, A., Chakraborty, P., Sha, S., Khatun, M., Hasan, M. and Banerjee, K. (2020). "Improvised Technique for Analyzing Data and Detecting Terrorist Attack Using Machine Learning Approach Based on Twitter Data", *Journal of Computer and Communications*, 8, 50–62. https://doi.org/10.4236/jcc.2020.87005

CHAPTER 2

5G Evolution and Revolution

A Study

Namita K. Shinde, Chetan More,
Payal Kadam and Vinod Patil

2.1 INTRODUCTION

Mobile communication is the use of technology that allows for information transfer without the need for a physical link, simplifying our lives and reducing our work and time. Current wireless communications systems have undergone a significant development over the past several decades. Each generation differs from the one before it in regard to standards, features, methodologies and functionalities.

2.1.1 First-Generation (1G) Technology

The advanced mobile phone system, or first-generation wireless mobile communication system, is an analogue technology created in 1980 and utilized for voice services. The development of cellular technology and the division of the land into discrete areas known as cells marked 1G's primary distinction from earlier wireless communication systems. Every cell featured a base station that communicated with mobile devices using radio waves and a transceiver as shown in Figure 2.1. Base stations had telephone network connections and made voice calls using frequency modulation techniques [1].

2.1.2 Second-Generation (2G) Technology

The first analogue mobile communication systems' mobile communication standards paved the way for the digital mobile telecommunications communication technologies of 2G. With a speed of 64 kbps, the 2G wireless mobile communication system is a digital innovation from the late 1980s that employs digital signals for speech transmission [2]. The GSM, TDMA and CDMA are the three main technology standards for 2G wireless connections, which are intended to replace analogue with digital technology. The most popular 2G mobile technology, GSM, offers worldwide roaming, allowing customers to use their mobile phone connections with higher quality and capacity across many nations [3]. The BSS, NSS and Operational Support Subsystem are the three main interrelated subsystems that make up the GSM architecture, as shown in Figure 2.2.

DOI: 10.1201/9781032699806-2

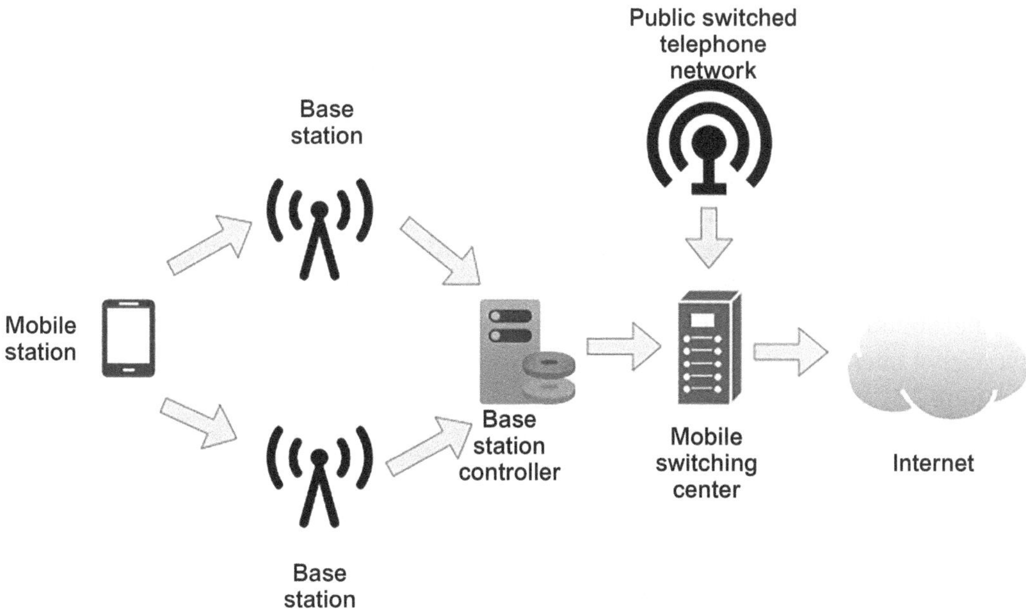

FIGURE 2.1 Architecture of 1G technology.

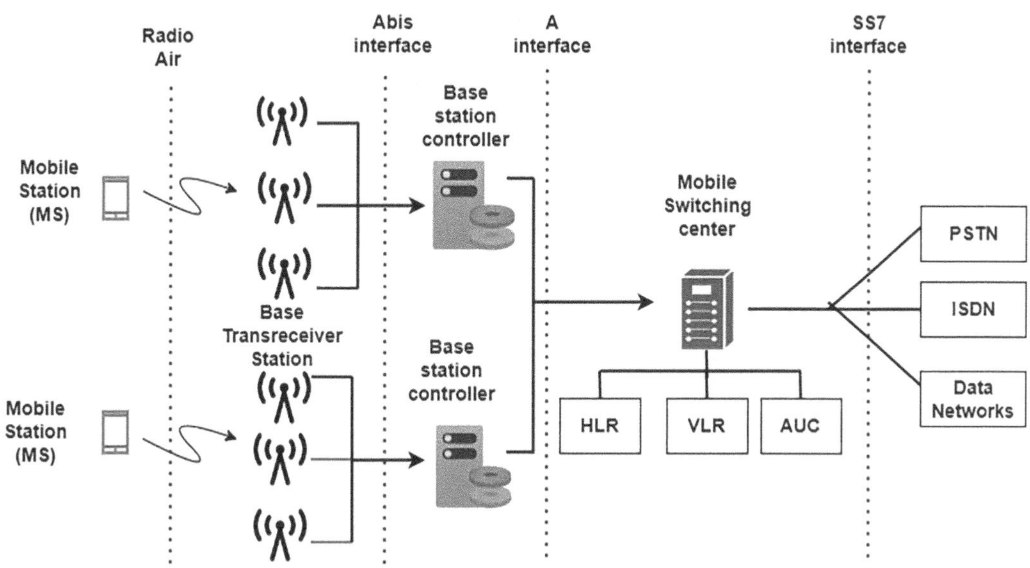

FIGURE 2.2 GSM system architecture.

Each BSS is made up of a large number of base station controllers (BSCs), which use MSCs to link the MS and NSS. By connecting MSCs to PSTN along with ISDN, whereas NSS manages system switching [4]. The GSM system's operation and upkeep are supported by the OSS, which enables engineers to keep track of, analyse and troubleshoot all its components [5].

2.1.3 Third-Generation (3G) Technology

After moving from second- to third-generation framework, users saw higher system and download rates, enabling uninterrupted video calls. The 3G core network has four key tasks to perform as shown in Figure 2.3. Circuit switching utilizes a network, in which a devoted connection is made available to a group of users during a specific time. Packet-switching makes use of an IP network, where IPs oversee transferring data between multiple devices. Mobility and session management, cost and connection with other network areas are all provided by SGSN [6]. The extremely sophisticated GGSN router manages internal communications between UMTS and other packet switched networks which are not part of internal system. The 3G core network has four key tasks to perform as shown in Figure 2.3.

2.1.4 Fourth-Generation (4G) Technology

Mobile services that include fast connectivity, Internet Protocol mobility, smart nodes and WWW-like services are made available by fourth-generation technology. The 4G standard, which only applies to mobile devices, has an information rate improvement from 20 to 60 Mbps and relies on LTE and WiMAX technologies. The 4G network is a fully IP-based mobile system and supports ad hoc and multi-hop network with improved spectrum efficiency [7].

Three essential elements make up the high-level network architecture of LTE: User Equipment (UE), Evolved UMTS Terrestrial Radio Access Network (E-UTRAN) and Evolved Packet Core (EPC) as shown Figure 2.4. The information bits are terminated in terminal equipment, and mobile termination manages all communication tasks. The numerous data packets that are transmitted between mobile devices and the core network are managed by an evolved packet core. Radio connection between end devices and the EPC is managed by E-UTRAN. One cell and one ground station can be connected simultaneously by LTE mobile. EPC connects with IP multimedia subsystem, internal and external packet data networks and other systems [8].

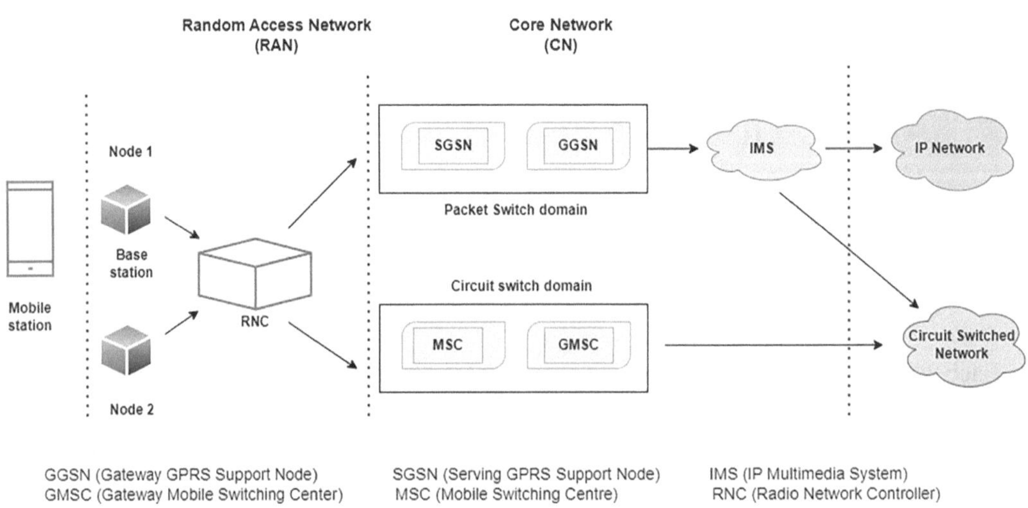

FIGURE 2.3 3G core network.

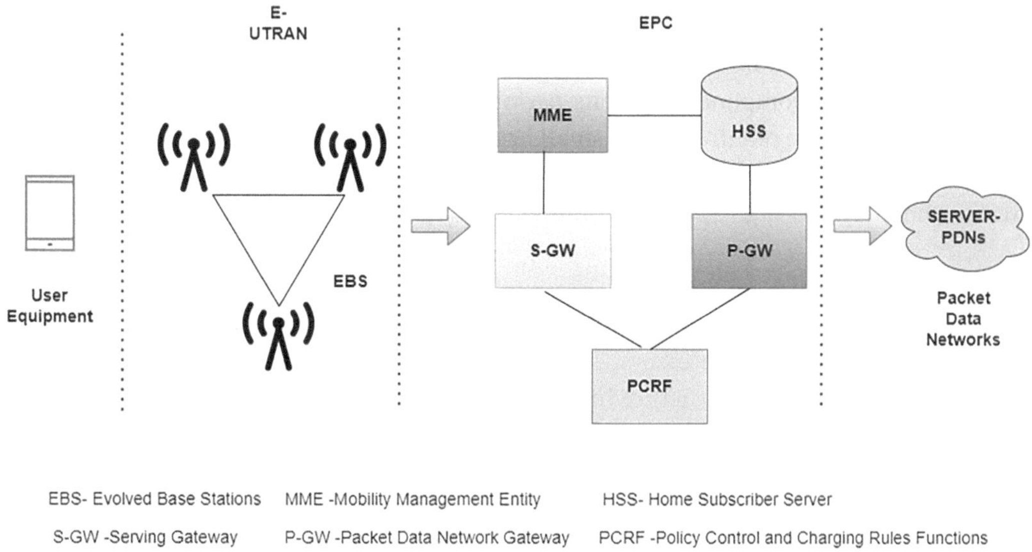

FIGURE 2.4 4G network architecture.

2.1.5 Difference between Different Wireless Technologies

The objective of offering reliable, high-quality communication comparable to wired communication is advanced significantly with every new generation of wireless communication. Mobile communication has become more commonplace recently as a result of the quick development of mobile technology. Table 2.1 illustrates the differences between each evolved generation.

TABLE 2.1 Differentiation between Different Wireless Technologies

Parameter	1G	2G	3G	4G
Development	1980	1990	2000	2010
Data Bandwidth	2 Kbps	14.4–64 Kbps	2 Mbps	200 Mbps to 1 Gbps
Technology	AMPS, NMT, TACS	GSM	WCDMA	LTE, WiMax
Frequency	30 KHz	1.8 GHz	1.6–2 GHz	2–8 GHz
Characteristics	Voice only	Data, voice, MMS, web browsing	Universal access portability, video calling	HD streaming, worldwide roaming
Core network	PSTN	PSTN	Packet network	Internet
Multiplexing	FDMA	TDMA, CDMA	CDMA	CDMA
Switching	Circuit	Circuit, packet	Packet	Packet
Examples				

2.2 UNIQUE 5G COMPONENTS COMPARED TO 4G

This section provides unique changes required in 5G communication technology, which can distinguish 5G components and its performance over 4G communication network. The study includes the requirement of small hexagonal cell structure for 5G tower antenna, beam-like spectrum formation technique in 5G tower antenna and mobile edge computing (MEC) techniques for high-speed communication [9].

2.2.1 Small Hexagonal Cell Structure

One of the primary distinctions between 4G and 5G is the most basic base station required to send signals. Like its predecessors, 4G transmits signals. Carriers will set up high-band 5G in small hexagonal cell structure in various locations since 5G uses small cell technology and operates on mmWave (millimetre wave) frequency bands, which are faster [10]. For its higher-speed and lower-frequency spectral response, 5G will continue to employ phone towers. Carriers are required to install tiny cells across various places due to the mmWave frequency. Even while mmWave technology now runs at a frequency higher than cellular technology, its broadcasts are weaker and cover less space. In 5G-capable areas, small hexagonal cell structure sites must be constantly erected to guarantee that the signals reach clients and companies. Small cells are radio access nodes for cellular networks that operate in the range of 10 metres to a few kilometres. Small hexagonal cell structure is important for the creation of a 5G communication wireless system/network. Coverage of tiny areas is provided by low-power base stations, or "small cells". Small hexagonal cell structure is somewhat like the cells that were previously used in various wireless networks. They can function at high data speeds while consuming little power is one of these cells' benefits. Smaller units in communication tower support the development of the 5G network with their ultra-high-speed and low-latency connection. To transmit data at fast rates, tiny cells in the 5G network require new technologies including MIMO (multiple input and multiple output), beam-like spectrum formation technique in 5G tower antenna, and mmWave. Hardware for small cells is extremely straightforwardly designed, making it simpler and quicker to implement. Small cell towers come in three different varieties: femto-, pico- and microcells

In the 5G communication spectrum frequency range is between 30 and 300 GHz, mmWave, and it is a very high communication frequency band. It enables highly fast wireless communication because it is a much less crowded portion of the spectrum. For the upcoming mobile networks, mmWave offers an extremely wide bandwidth. MmWave has many benefits, but it also has some drawbacks. For example, because mmWave signals are extremely high-frequency, they collide more frequently with airborne objects and lose energy more quickly. Signals from mmWave are also obstructed by structures and trees, which reduces their range. Numerous tiny cell stations are constructed to bridge the distance between the end user and ground station in order to address these problems. Because a tiny cell has a much smaller coverage area, the installation of one is dependent on the local population. In a densely populated location, the typical small cell distance varies from 10 to 90 metres. Many of the writers in the study combined the usage of small hexagonal cell structure for 5G tower antenna and massive MIMO. They also analysed a variety of 5G technologies, the beam-like spectrum formation technique in 5G tower antenna, massive MIMO with non-orthogonal multiple access (NOMA), and one-device-to-another-device (point-to-point) (D2D) communication. The control of interference, spectrum efficiency, resource management, energy efficiency and industry are just a few of the issues that are covered. Additionally, the author provided a thorough explanation of all the problems encountered while integrating tiny cells with different 5G technologies [11].

There is discussion of several issues, including congestion control, spectral efficiency, management systems, energy efficiency and backhauling. The author also provided a comprehensive overview of all the problems encountered while integrating tiny cells with different 5G technologies. As shown in Figure 2.5, mmWave has a better range, obstructions can quickly block it. One of the main issues with mmWave signal transmission is this. The small cell could be positioned close together to readily transmit the signals in order to address this problem.

The 5G small cell structure allows for an increase in cell density and network capacity. Despite the identical promises made by 4G, 5G will ideally fill the holes left by 4G as the latter never fully met its ambitious general speed objectives. The capacity of smartphones and connections will increase with the advent of 5G as channels get denser and can support more users and connected devices [12]. Despite the promised advancements of 5G, its promises won't immediately come to pass. Operators will still need time to address any problems and irregularities that 5G may introduce.

2.2.2 Beam Spectrum Formation Technique in 5G Tower Antenna

A crucial component of wireless networks, beamforming directs the signal transmission of the signals. An effective wireless connection is made toward the receiving end using the 5G beam spectrum formation technique in the 5G tower antenna. Moving messages to certain locations is particularly challenging when hexagonal cell structure is not employing beamforming. Small cells can transmit signals in a specific direction to devices like mobile phones, laptops, autonomous vehicles and Internet of things (IOT) devices utilizing beamforming to address this problem. The 5G network is becoming more energy-efficient thanks to beamforming. In general, there are three types of beamforming: hybrid beamforming,

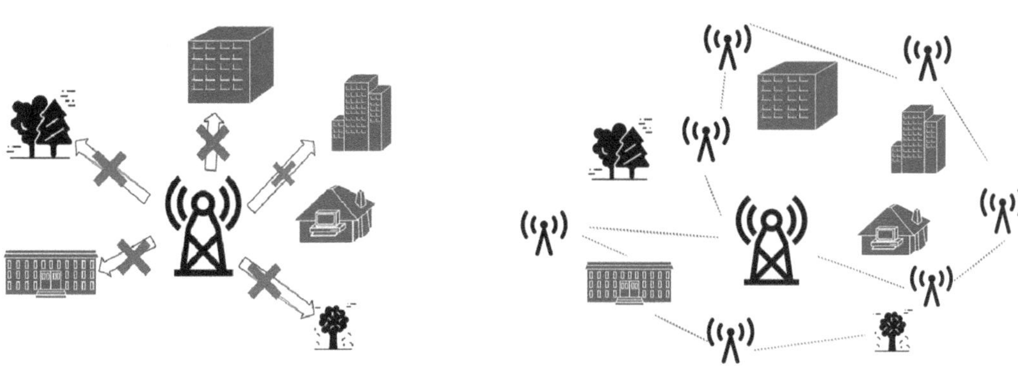

FIGURE 2.5 Representation of communication with and without small hexagonal cell structure.

analogue beamforming and digital beamforming. Multi-user MIMO is equivalent to digital beamforming, which is mostly utilized in LTE Advanced Pro and 5G New Radio (NR), for it is the international standard for a more capable and unified 5G wireless air interface. The digital beam spectrum formation technique in 5G tower antenna increases the cell capacity of wireless networks by allowing the data to be transmitted to numerous users simultaneously using the same frequency or time resources. Analogue beam spectrum formation technique in tower antenna [13]. 5G NR analogue beamforming, which increases coverage in the mmWave frequency range, is a very significant method. Because only one beam is created per set of antennae in digital beamforming, there is a potential of substantial pathloss in the mmWave. High pathloss in mmWave is reduced through analogue beamforming. By combining analogue and digital beamforming techniques, hybrid beamforming is created. Hybrid beamforming will be employed while implementing mmWave in the 5G network.

Nature is not omnidirectional, and the wireless signals of the 4G network are spread out across a large region. Energy is quickly used up as a result, and users seeking to get these signals have interference problems. To solve this issue, the 5G network employs the beamforming technique. Signals used in beamforming are directed. Signals appear to be moving in an unseen cable because they appear to move like something of a laser beam from of the base station toward the user. Beamforming aids in achieving a higher data rate; since the signals are directed, it uses less energy and reduces interference. Develop various methods that improve the 5G mobile network's system performance and eliminate interference [14]. The writers of this survey study discussed the many difficulties encountered when creating an algorithm for efficient beamforming, primarily concentrated on several design parameters, such as power consumption and performance assessment. Additionally, they discussed several beamforming-related problems, including CSI, computing efficiency and antenna correlation. Additionally, they discussed different studies that show how beamforming facilitates the use of MIMO in next-generation cellular operators. Figure 2.6 shows the visual representation of beam spectrum formation technique– and non-beam spectrum formation technique–based communication.

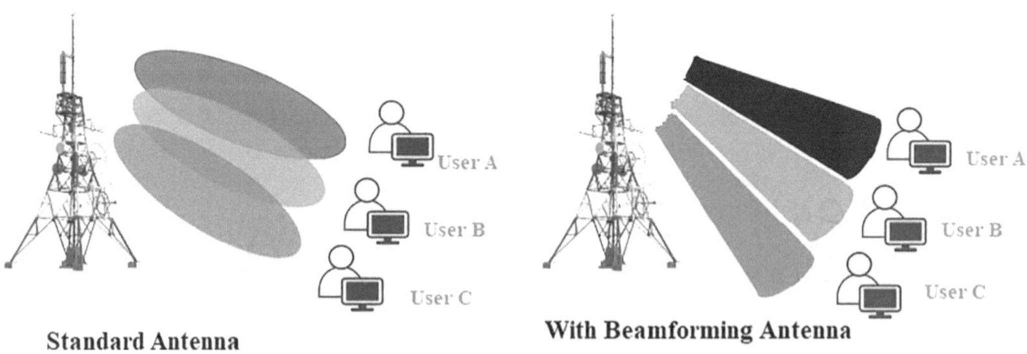

FIGURE 2.6 The visual representation of non-beamforming and beamforming-based communication.

- **Latency:** In terms of key differences, latency separates 4G and 5G. 5G promises to have low latency of less than 5 ms, while 4G latency is often between 60 and 98 ms. Furthermore, as latency continues to decrease, improvements in other areas, including faster download rates, take place.
- **Potential download speeds:** While 4G made several VoIP opportunities possible, 5G builds on and improves past claims of high theoretical download speed. The fastest download rate for 5G is 10 times faster than the fastest download speed for 4G, which was 1 Gbps.

2.2.3 Mobile Edge Computing (MEC)

By bringing resources in the cloud are closer to the end user, mobile cloud technology is an expanded kind of cloud computing. When we discuss computers, cloud computing is the very first concept that comes to mind. A highly well-known technology, cloud computing gives end consumer's access to a variety of services. Cloud computing, however, has a lot of drawbacks. Because of the latency caused by the cloud services' distance from end users and the necessity of downloading the entire application before usage, the device is also put under more strain [15].

Mobile edge computing, which creates an edge between the end user and cloud server, brings cloud computing closer to the end user. This edge currently offers every service, including virtual software, video conferencing and other services, which increases the effectiveness of cloud computing. The application is divided into two parts, the first of which is accessible on a cloud server and the second of which is on the user's device. This is another essential component of mobile edge computing. Therefore, the user of the programmer that improves a user's device's performance does not need to download the complete application on his device. Additionally, cloud services with very low latency and bandwidth are available through mobile edge computing. The 5G network's overall performance is improved by the successful implementation of mobile edge computing. It shows how cloud computing with mobile edge computing vary graphically in Figure 2.7

2.3 5G SPECTRUM

After the first complete collection of 5G specifications was completed, the commercial use of 5G began in 2019. With its innovative wireless technology standards, 5G ushers in a new era of global digitalization. This is especially true in terms of data transmission rate, latency, accessibility, portability and even the number of devices. Users of 5G can anticipate considerable advancements, including data transfer rates of up to 10 Gbps, substantially reduced latency (over almost 10 ms at a higher capacity), reliability and quality of service. The capabilities of 5G have really distinguished this cutting-edge mobile network from its forerunners. 5G is the first service to use the mmWave spectrum, a revolutionary tech of frequency band, in contrast to its forerunners. Additionally, 5G is the key IOT infrastructure and can incorporate a wide range of new technologies, including massive (MIMO), communication-based software-defined networking (SDN) and device-to-device (D2D) communication. 5G represents a dramatic advance in communication networks. 5G delivers a range of applications using a single platform, including improved broadband Internet

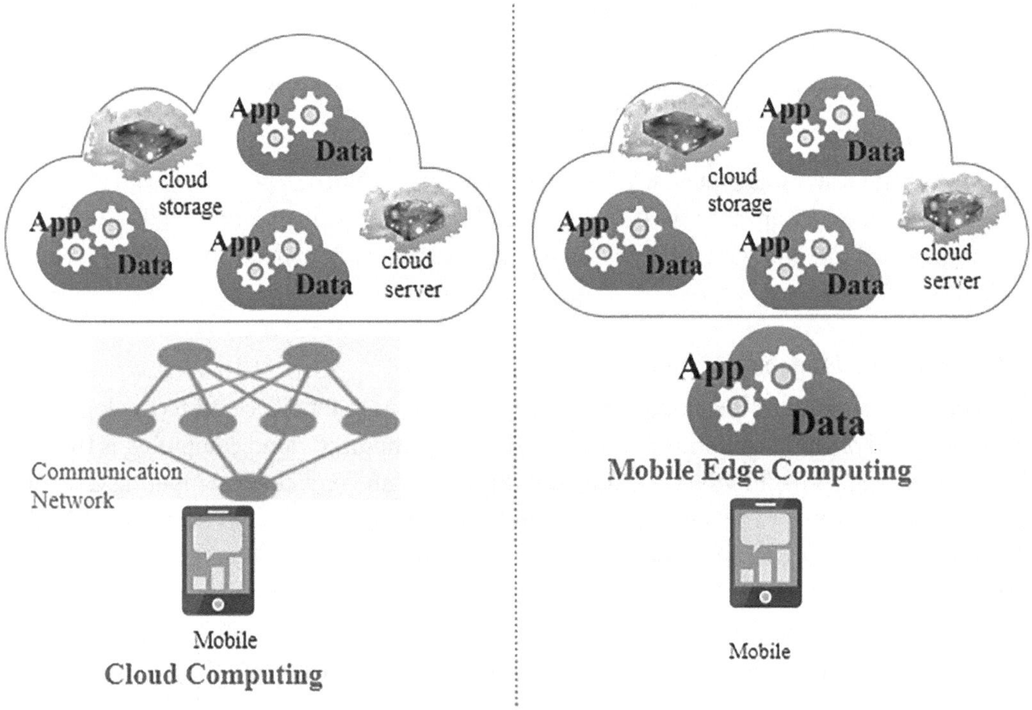

FIGURE 2.7 Cloud and edge computing on mobile devices visually varying.

communications, automated driving, virtual reality and the IOT. Going beyond 5G is necessary to match the escalating technical demands and requirements at all levels due to the recent decade's rapid technological breakthroughs and inventions [16].

2.3.1 Spectrum Operation

Spectrum is a key component of wireless networks. It serves as the "airwaves" that carry our daily communication services, including TV, Wi-Fi and mobile. The variety of 5G applications and services previously stated will call for a variety of spectrum bands with various properties, serving various needs and incorporating both minimum and maximum frequencies:

A. To enable 5G coverage across large areas, a spectrum at lower frequency region is needed, especially below 1 GHz.

B. Below 6 GHz spectrum at very huge frequency region with comparatively big bandwidths is needed to accommodate a very high number of linked System and devices and to let them to operate at higher speeds.

C. The mmWave spectrum, which works at very high frequencies exceeding 24 GHz, offers extremely large bandwidths and ultra-high capacity. The rollout of 5G technology in mmWave bands is probably going to start off with a focus on areas with high bandwidth to manage traffic demand, or perhaps to specific places or premises needing services with considerably bigger efficiency and/or peak data rates. Cells at these frequency components will likely have smaller antennas and a reduced cover radius (between 50 and 200 m) (Gbps).

LTE (4G) bands that have already been harmonized below 1 GHz, especially the 700 MHz and 800 MHz band, will be crucial in offering ubiquitous coverage in next-generation (5G) or pre-5G networks. This will enable nationwide and indoor 5G coverage. According to the authors, the 450 MHz band, where standardized LTE usage criteria are now being defined at CEPT, will also play a vital role in providing comprehensive coverage for solutions and services for next-generation mobile wireless communications. The 450 MHz frequency range has set conditions for LTE operation. The frequency band between 3400 and 3800 MHz, which is already synchronized for mobile networks and contains up to 400 MHz of spectrum [17]. 5G communication cell structure and band sharing is depicted in Figure 2.8.

2.3.2 Frequency Ranges in 5G

A. To narrow the digital divide and improve broadband connectivity between urban as well as rural areas, more low-band (i.e., 1 GHz) bandwidth capacity will indeed be needed. Additionally, it will enhance global in-building 5G services and promote the expansion of the IOT. All current low-band frequencies (900 MHz, 850/800 MHz and

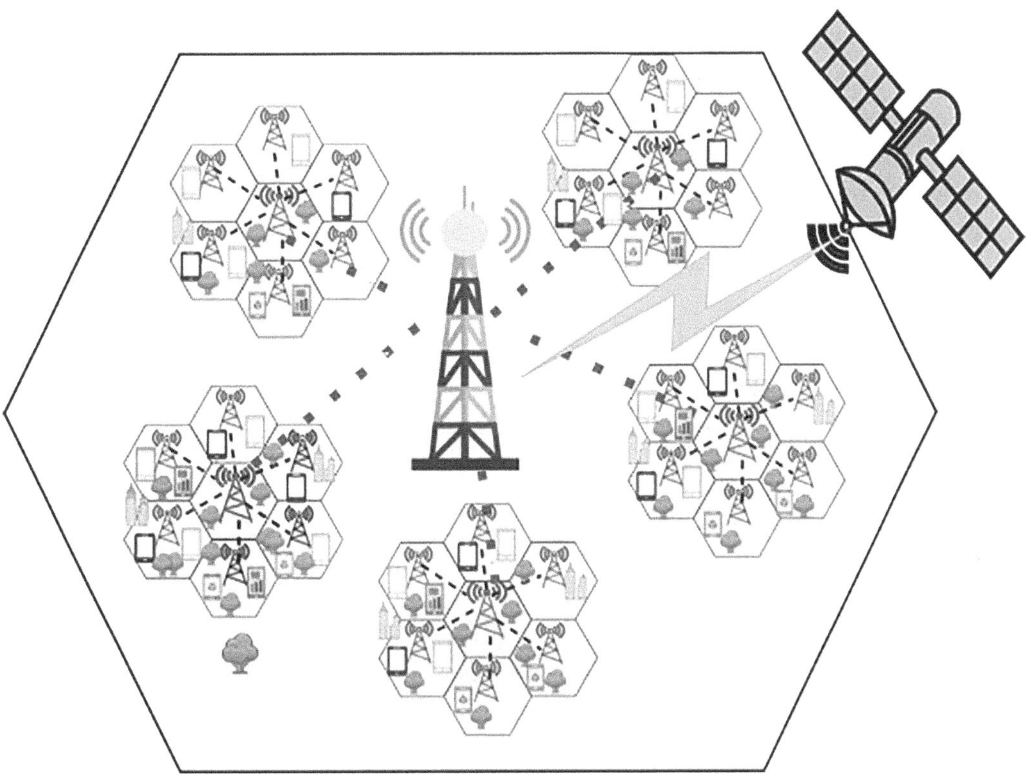

FIGURE 2.8 5G communication cell structure and band sharing.

700 MHz) must be reserved for mobile phones. Support for further lower frequency bands is gaining traction, with interest in 600 MHz in particular.

B. For 5G services, mid-bands often provide a decent balance of coverage and capacity. The 3.5 GHz band should have as much continuous spectrum as possible assigned by regulators (3.3–4.2 GHz). Operators should also be granted licenses for the usage of the 2.3 GHz as well as 2.6 GHz channels for 5G.

C. 5G technologies like ultra-highspeed mobile broadband require high bands. Without mmWave spectrum, 5G wouldn't be able to provide the best data speeds. It is crucial that governments grant broadly standardized spectrum, such as at frequencies of 26 GHz, 24 GHz, 32 GHz, 28 GHz and 40 GHz, as all bands may be readily supported by a variety of reasonably priced devices with minimal complexity. Radio spectrum for 5G and its uses are shown in Figure 2.9.

2.3.3 Effect of Geographical Regions in 5G

Without the guarantees of long-term, dependable, predictable spectrum access, the risks associated with network investment are greatly enhanced. Wider coverage regions and higher levels of service assurance provided by 5G spectrum have been essential to the massive global growth of mobile technologies.

Substantial amounts of exclusively licensed spectrum that span entire nations are available for 5G services. To meet a variety of use cases, including those in mmWave bands and mid-bands, national mobile spectrum licenses are still crucial. Addressing high-capacity requirements in hotspots and fixing Wi-Fi network in cities, suburbs and rural towns, as well as other regions with high population densities of usage that can be in a broad range of locales are some of these (e.g., factories, parks, train/bus stations, etc.). The usage of 5G access bands for backhaul in diverse locations is also possible. To encourage operators to use their frequency or make it accessible to others in cases when it won't be used or won't be utilized in a reasonable period of time, regulators can establish license conditions and duties. The mmWave frequency bands utilized by the mobile industry may

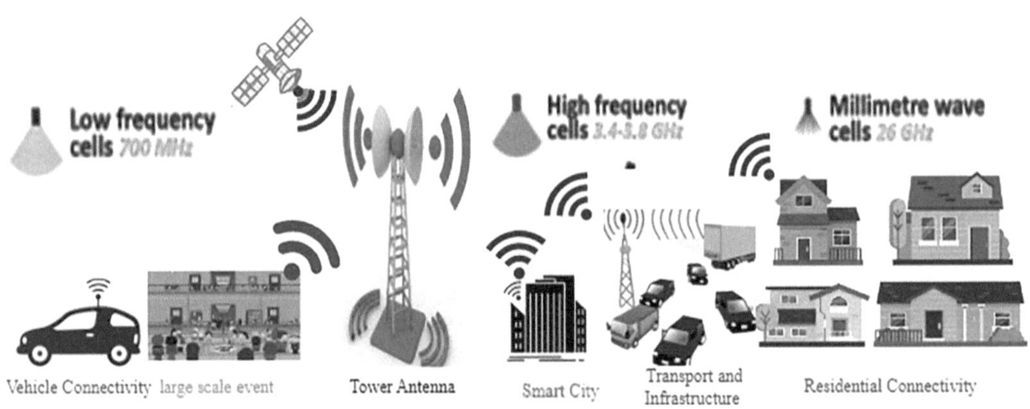

FIGURE 2.9 Radio spectrum for 5G and its uses.

include exclusive broad area licenses in strong market demand locations (such city centres) and local licensing in all other locations. It should be highlighted, though, that managing such approaches is substantially more difficult for regulators, as they must make sure the various possible assignments are fair, hoarding/underuse is watched for and avoided, and interference issues are handled. Local licensing can also make it more difficult for operators to implement new technologies because it makes it harder to predict when they will have access to enough spectrum and because it may place restrictions on use (such as power caps) to prevent interference with nearby licensees [18].

2.3.4 Sharing mmWave Spectrum

Wireless local area networks (WLANs) and mobile networks (5G and beyond) are both utilizing mmWave (mmWave) communications as a significant disruptive technology (802.11ad and beyond). mmWave frequencies offer orders of magnitude higher bandwidths, although spectrum availability is constrained in conventional bands below 6 GHz. In addition, transmissions with extremely narrow beams are a common feature of mmWave communication, allowing for even more benefits from direction isolation between mobiles. A larger peak each data rate, a high traffic density and extremely low latency are just a few of the audacious 5G requirements that can be met by mmWave thanks to their vast spectrum and geographical degrees of freedom. In contrast to lower frequencies, mmWave bands offer a variety of special advantages for 5G [19].

2.3.5 Mandatory Beamforming Requirements

Beamforming is a need in all systems running at mmWave frequencies to account for the noticeably greater pathloss at these frequencies. For instance, the IEEE 802.11 specification allows for up to four broadcast and receive antennas as well as 128 sectors. In 802.11ad, beam shaping is required and is supported on both the transmitter-side and receiver-side. Additionally, the 3GPP is anticipated to finalize the 5G beamforming specifications as part of the 5G NR construction task. As a result, beams offer a new dimension for spectrum sharing across various access modalities.

2.3.6 Potential for "Unlimited" Spatial Reuse

The foundation of cellular communications is spatial radio spectrum reuse, which is already used by wireless communication systems. The footprint of disturbance from each communication path can be made to resemble more closely a line than an area in mmWave systems that employ both transmit-side and begin receiving beamforming, pushing spatial spectrum reuse even more in the direction of one dimension. This would permit indefinite spectrum utilization of spectrum in the ideal situation of ultra-narrow beams.

2.3.7 Satellite Services

Geostationary communications satellites that, for example, supply broadcasting feeds to TV channels, radio stations and broadcast networks are officially referred to as FSS (fixed satellite service). The FSS uplink, which connects the FSS to a satellite, is allotted in the 27.5–30 GHz range, which is close to the 24.2–27.5 GHz spectrum designated for 5G. As a result, interference from neighbouring channels could cause problems when 5G and FSS are used together. To enhance the coexistence of 5G and FSS, a variety of cognitive approaches can be used. There have been relatively few studies on the cohabitation of FSSs with mobile communication base stations in the mmWave frequencies, mostly theoretical ones.

Important new considerations include how effectively the interference levels may be reduced by 5G mmWave devices using a variety of antenna topologies and examining the overall interference brought on by the widespread use of 5G systems on uplink FSS. Studies in, performed in the worst-case scenario of co-channel cooperation, have recommended that spatial sharing is significantly more practicable than the scenario of IMT-advanced systems. This is because beamforming technology is used and because the links in mmWave frequency are relatively short. Even in this most extreme example, the necessary safety distances surrounding FSS are far smaller (1 km as opposed to hundreds of kilometres) than those previously suggested. By utilizing collaboration among many 5G BS, more spectrum access advancements may be accomplished. According to these findings, the 5G mmWave network's capacity to offer coverage may be affected by the presence of highly directed FSS transmission. However, the outage zone is confined because to the highly directed FSS transmission (as illustrated in Figure 2.10), and combining null forming at 5G users and collaboration by numerous BS to increase signal power at the impacted UE may help to mitigate its effects [20].

2.3.8 Setting Spectrum

The success of 5G as a whole is threatened by spectrum that has been set aside for local use, especially by vertical businesses in priority 5G channels (i.e., 3.5/26/28 GHz). To enable mobile carriers to provide the quickest 5G services, set aides might restrict the allocation of reasonably large contiguous blocks. By giving some users preferential access to the

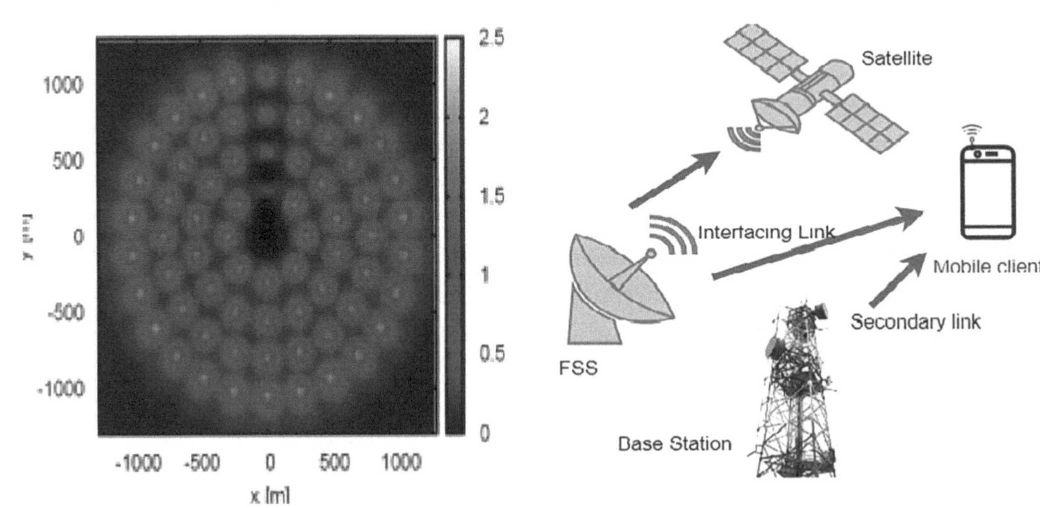

FIGURE 2.10 Hybrid communication system with signal intensity distribution. (Left panel: signal intensity distribution plot; Right panel: schematic diagram of hybrid communication system.)

spectrum rather than allowing them to compete for prizes, they can also impede equitable access to the spectrum. Set asides can produce artificial scarcity, which runs the danger of driving up the price of spectrum, which is associated with slower rollouts, less coverage and slower speeds.

Regulators should particularly avoid setting aside spectrum when doing so will prevent them from achieving the goal of having 100 MHz per operator accessible in priority mid-bands (such as 3.5 GHz) as well as 800 MHz for mmWaves (e.g., 26 or 28 GHz). Regulators should be aware that additional mid-band and mmWave spectrum may be required for commercial 5G technology over a long-time frame (for example, 2025–2030) as consumption increases, even if they are able to reach these requirements in the short term. Setting aides for constrained use cases often can result in ineffective spectrum utilization.

National set-asides, for instance, are likely to go underutilized in many places since vertical businesses are unlikely to utilize bandwidth in key 5G bands very extensively throughout any country. Instead, mobile carriers may provide specialized 5G services for industry verticals, which can then be gained from network slicing, tiny cells and broader geographic coverage. Strong advantages for verticals are also provided by the bigger and more varied spectrum holdings and deployment expertise of mobile carriers. Regulators would be wise to allow the set-aside spectrum to be made accessible through market processes after a reasonable period should this turn out to be the case given the substantial danger that it may go underutilized in many locations (like sunset clause) [21].

To demonstrate that a set-aside strategy would produce a better socioeconomic result for a nation than a totally market-based award, a thorough expense analysis should be carried out. If a set-aside is adopted, authorities should take precautions to limit the damages to public 5G services, consumers and companies that depend on them (for further details, please consult the GSMA document “Mobile Communications for Industry Verticals: Bandwidth Best Practice”). Approaches to voluntary spectrum sharing are often preferred to set aside. They may be utilized to support all possible 5G users, including verticals, without restricting the amount of spectrum that is available for public 5G services.

For instance, MNOs may be allowed to license their spectrum holdings to verticals so they could create their own personal 5G networks if they so desired. Instead, in Finland, the regulatory body imposed a requirement on telecommunications company licensees operating in the 3.5 GHz band to rent their spectrum to industries where they could not offer adequate services. In a rising number of nations, like Sweden, where frequency band has been subleased to a specialized vertical micro-operator, mobile providers are pooling their spectrum with several other customers [22].

2.4 BASIC NETWORK

We will outline the 5G core architecture and define its components in this part. We will also contrast the 5G design with the existing 4G architecture.

2.4.1 4G Architecture

The network architecture had relatively minor incremental modifications as 4G developed from its 3G predecessor. The essential elements of a 4G core network are shown in the schematic of the 4G network architecture in Figure 2.11.

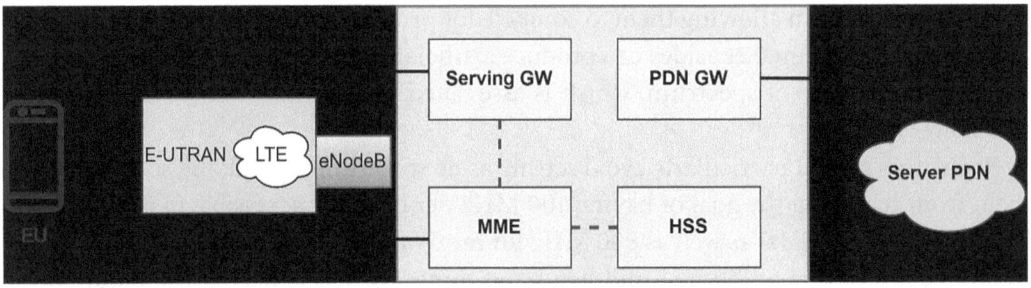

FIGURE 2.11 Architecture of 4G.

The LTE radio access network in the 4G network architecture connects user equipment, such as smartphones or cellular devices, to the evolved packet core and eventually to external networks, such as the Internet. The user plane (user data traffic) and the control plane (network management data traffic) are separated by the Evolved Node B (eNodeB) and sent separately into the EPC [23].

2.4.2 The 5G Network Architecture Diagram Illustrates How These Components Are Associated

In 5G, which was designed from the bottom up, network functions are broken down by service. Because of this, this architecture is also referred to as the 5G core service-based architecture. Figure 2.12 shows schematic of a 5G network topology displays the essential elements of a 5G core network [24].

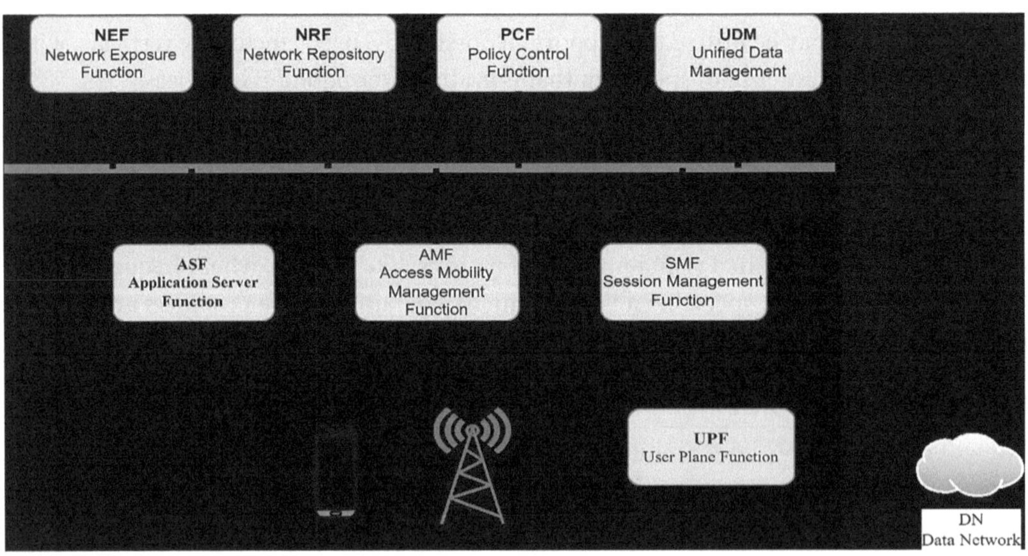

FIGURE 2.12 Architecture of 5G.

2.4.2.1 5G Core Components

A. *User Plane Function (UPF)* is one of the key components of a 5G core system design. It performs the most important function in data transfer and is a primary network function (NF) of the 5G core network (5GC). In the 5G architecture, it connects to the data network (DN). Additionally, it handles packet inspections, QoS (quality of service) handling and acts as an anchor point for intra and inter-RAT mobility, with additional responsibilities in the horizon. A high-performance forwarding engine for user traffic is provided by UPF, a virtual network function (VNF). While maintaining compatibility with all user plane capabilities, it achieves ultra-fast packet forwarding using vector packet processing (VPP) technology. As a result, packet processing and traffic aggregation may be done more closely to the network edge, making UPF the key enabler of innovative 5G applications and customer-specific edge services.

B. *Radio Access Network (RAN)* provides a path where the User Equipment (UE) can connect to the Access Mobility Management Function (AMF) to establish the wireless link to the core network. RAN is a very important element since it is in charge of the allocation of radio resources while supporting the call from the core network as well as catering for mobility of user terminals.

C. *Data Network (DN)*, e.g., operator services, Internet access or third-party services, Internet connectivity, operator services or other services.

D. *Session Management Function (SMF)* interactions with the decoupled data plane, PDU session management with the user plane function, and PDU creation, update and removal are its main responsibilities (UPF). The SMF must be wholly developed and supplied as a cloud native network function that can be dynamically launched and scaled-up on demand in an automated manner in order to comply with the architectural requirements of 5G. When it comes to high-availability control components that have asynchronous call flows across geographically diverse infrastructures and necessitate long- and short-lived state maintenance for sessions traversing elements that might quiesce without warning, this is a particularly challenging proposition. Therefore, these services must use tried-and-true design patterns for creating and deploying massively scaled web applications while simultaneously adhering to the limitations of real-time computing.

E. *Access Mobility Management Function (AMF)* is responsible for adhering to the termination of the NAS (NG1), NAS ciphering and NAS integrity protection; termination of the RAN Control Plane interface (NG2); mobility management; and lawful intercept (for AMF events and interface to lawful interception system).

F. *Application Server Function(ASF)* is connected with the Access Mobility Management Function and offer services to the User Equipment through the Radio Access Network. ASF acts as an interface for performing application specific tasks and for communication between the applications and services to ensure smooth service delivery and management.

G. *Network Exposure Function (NEF)* enables easy access to network capabilities, fostering innovation and collaboration among partners and customer. Customers can securely expose their services and capabilities that are offered by 3GPP 5G network functions using NEF, which is closely tied to the 3GPP 5G Architecture. These include third-party functions and internal exposure or re-exposure. Due to the data-richness and resourcefulness of NEF's apps, network ecosystems can be connected to them. This implies that enterprise applications enhance network data and resource access for various ecosystems. An NEF function enables data transfer within signaling messages via a control plane, doing away with the pricey user plane connection.

H. *NF Repository Function (NRF)* supports the function of service discovery. As a result, it can accept NF Discovery Requests from NF instances and can provide details about NF instances that have been found.

I. *Policy Control function (PCF)* supports a uniform policy framework to control network behavior by providing: Policy rules for the function(s) of the plane that apply them.

J. *Unified Data Management (UDM)* supports the following functions: Authentication Credential Repository and Processing Function (ARPF); this function keeps track of the long-term security credentials required for AKA authentication.

2.4.2.2 Adoption of 5G Geographical Architecture

A global step function in 5G integration for diverse geographical locations is necessary due to the infrastructure required for standalone 5G deployment. While other nations throughout the world lag far behind, technology-leading regions like North America, Asia and Europe are accelerating implementation. By 2025, there are expected to be more than 2 billion mobile 5G connections, with around 200 5G networks currently operational throughout the world.

The expansion has been more difficult in Europe because of the close proximity to neighbouring nations and the abundance of carriers. The European Commission has started a guideline known as the Digital Compass that calls for 5G access in all populated areas by 2030, even if adoption has lagged in other countries.

2.4.2.3 Current 5G Deployment Situation

China, Japan and India are among the industrialized nations that have made significant financial and practical investments in the 5G changeover. For the global electronics and software design and production businesses, new antenna, infrastructure hardware and software technologies are a gold mine with an emphasis on quick deployment. Following a spectrum auction in August 2022, the four main telecom companies in India are launching 5G services, and China is anticipated to follow [26].

2.4.2.4 The 5G Architecture's Security

A wide range of applications are made possible by the extensive use of cloud-based resources, automation, network slicing and other cutting-edge innovations in 5G implementation. These changes add new security risks and expose more "threat vectors" inside the 5G security infrastructure. Although 5G security protocols draw on past mobile technology iterations, the "trust model" has been expanded to encompass more parties involved in the service delivery chain. IOT and user growth are producing an increasingly higher number of endpoints, and a lot of these network inputs are no longer being monitored by humans. Public key-based encryption methods to reduce the possibility of information manipulation and unified authentication to detach authentication from access points are both part of the 3GPP standards for improved 5G security.

As 5G critical performance nodes become more virtualized, ongoing security effectiveness monitoring and assessment are crucial. End-to-end 5G network security monitoring that covers the system architecture, devices and apps is among the best practices.

The exponential speed improvements that users have come to expect from each new generation of mobile networks are undoubtedly being delivered by 5G, but speed is only the beginning. Many people have nicknamed 5G the new Industrial Revolution due to the profound changes it will bring about in a variety of sectors, including personal transportation, manufacturing and agriculture. The multifaceted 5G architecture, which combines MEC, NFV huge MIMO and a core architecture centred on services that is aligned with the cloud, is at the centre of this paradigm change. The actual enablers of the upcoming 5G transition will be 5G test solutions built to support this architectural seed modification [27].

2.5 5G – TECHNOLOGY CHALLENGES

Challenges are a natural aspect of new developments, and 5G is no exception. It faces major difficulties. When we look at the development of radio communication, we can see how speedily it advanced. The journey only takes roughly 40 years to get from 1G to 5G (considering 1G in 1980s and 5G in 2020s) [28]. The main challenges we've encountered along the way have been a lack of infrastructure, poor methodological approaches and high costs. The most important questions on everyone's minds are "How far will 5G be viable?" because there are still dozens of nations that use 2G and 3G technologies and aren't even aware of 4G and "Will this only benefit some developed countries' technologies, or will it also help developing nations?"

2.5.1 Problems Associated with 5G

The 5G problems are divided into the two categories to help answer these questions:

1. **Bands of Frequencies:** Unlicensed airwaves at 6 GHz will be utilized by 5G. The 5G NR specification for 5G networks includes a variety of frequencies inside the sub-6

GHz frequency range, including mmWave of the RF spectrum. If 5G bandwidths across diverse places are not synchronized, cellular equipment manufacturers may run into problems.

2. **Massive Data Volume:** Advanced technologies cause every network's data volume to grow yearly, and this trend is accelerating. Each network must be able to manage massive amounts of data since several applications are capable of high-definition video calling, multimedia broadcasts, uploading, etc. Because of the increasing industry trend toward video standards, streaming video is far more searched than traditional message content. With technologies like audio-visual gaming, wearable technology and virtual reality to provide an enhanced user experience, high-speed networks are necessary.
3. **MIMO Technology:** Improved MIMO antenna arrays used throughout 5G-MIMO technology will provide quick data delivery to individual users. Raising the amount of radio transmitters at the base station and the mobile device is required for MIMO in order to maximize data flow by sending and receiving information concurrently (UE). Complex device abilities, as well as base station and device capabilities, are needed for MIMO technology [29].
4. **Beamforming:** Individual customers will enjoy quick data delivery because of the 5G-MIMO MIMO technology's sophisticated MIMO antenna arrays. To increase data transmission by communicating and receiving information in real time, MIMO necessitates the addition of transmitter only at base stations and mobile units. Advanced devices, base stations and user equipment capabilities are needed for MIMO technology.
5. **Communication between Devices:** By just using one smartphone as a central repository of information for that other equipment that are not able to access to base station signals, D2D communication, a revolutionary concept, seeks to enhance wireless services. Device-to-device communication is among the most effective ways to communicate in emergency situations (such those resulting from natural disasters) where connectivity is inadequate or non-existent. Complicated data transmission techniques, however, are required to set up D2D communication.
6. **Exceptionally Low Latency Service:** Mission-critical applications and self-driving cars require exceptionally low delay support to keep running well. Any delay in these applications could have unanticipated and severe consequences. Less than 1 ms of delay is required for medical applications like virtual surgery.
7. **Network with Ultra-Reliability:** A highly reliable network is required for emergency applications and services to quickly send out warnings in emergency situations. Wireless networks, among other things, are required for communication

between health monitoring equipment, virtual healthcare devices, rescue and fire systems, police and ambulance services. Real-time tracking of patients with specific needs (such as measurement of blood pressure, heartbeat and blood glucose) is increasing, and this trend is expected to continue. Communication between patients and doctors is essential for reporting, diagnosis and treatment. Most healthcare remote monitoring applications necessarily require a very stable network.

8. **Privacy and Security:** Safety is an essential component of any wireless transmission system. The 5G network must provide security and privacy to end users. Security is a difficult task to achieve due to the variety of technologies and the number of network-connected devices. End-to-end encryption techniques have been developed to ensure secure communication between devices and cloud-based application servers.
9. **Intelligent Automobile:** The smart vehicle industry is another driving force behind cutting-edge technology such as IOT, which uses a 5G network. Modern automobiles (vehicles) are including sophisticated multimedia devices that can connect to a mobile network. Smart automobiles will create future data communication hubs for several kinds of communication, which include vehicle communication, vehicle to infrastructure and vehicle to other devices. Autonomous driving requires networks with extremely low latency and reliability.

2.5.2 Infrastructure Requirements for 5G

Per the need of recent applications and data transmission requirements, the following are the infrastructure essentials for 5G:

2.5.2.1 Small Cells 1

The line-of-sight transmission path required for mmWave signal propagation makes it extremely susceptible to interference from nearby structures, vehicles and surroundings. Small cells, such as microcells or femtocells, will be used in 5G technology for transmission. By serving a limited number of users within each cell, tiny cells can give high data rates.

Small cell deployment across a wide geographic area requires a significant financial investment. Several deployments of micro cells, possibly even millions, are required to fully cover a big area or nation. Another difficulty is implementing a new system without affecting the LTE and LTE Advanced network. Due to the varied frequency spectrum, which calls for a different set wireless antenna systems, MIMO and beamforming capabilities, network improvement is frequently difficult.

There will be billions of connected devices by 2020, which is quicker growth than anticipated. By the year 2020, analysts predict that 50 billion smart sensors and devices will be linked to the Internet. Industries are growing smarter because of the use of sophisticated sensor technologies to collect vast volumes of data in real time. The Internet of things, which is frequently referred to as the next industrial revolution, helps businesses solve a range of problems by performing predictive analysis on data. The machine-to-machine (M2M) communication protocol was created to allow devices linked to a network for information gathering and information exchange [29].

2.5.2.2 IOT

Several wireless technologies, such as low voltage Bluetooth, Wi-Fi, LoRa WAN, ZigBee, etc., will be able to send data through the Internet of things. To support several technologies, gateways and networks must be interoperable and efficient enough to process massive amounts of data from millions of devices.

To securely transfer information between the nodes and the centralized data processing unit, advanced encryption algorithms are needed. Every year, more gadgets and sensors using different wireless standards hit the market. To support many future devices and technologies, networks must be scalable.

Because 5G is designed to work with a multitude of applications, its complexity is one of the main challenges to overcome. Because 5G offers a platform for several wireless technologies to coexist, technology suppliers must address challenges relating to signal spectrum, transmission techniques, security, network interoperability, etc. The tremendous need for a smarter network has accelerated the development of the 5G standard faster than expected. The first 5G network that is completely functional won't be available until 2020, according to analysts.

2.5.3 Advantages of 5G Technology

The benefits of a 5G cellular network are the reason that it is gradually gaining popularity around the world [17]:

A. **Superfast data transfer is possible with 5G:** The fastest benefit of the 5G broadband cellular type of network may be how quickly it operates. Data transfer rates of at least 1 Gbit/s (gigabits per second) are anticipated from 5G, dramatically accelerating Internet speed. Experts predict that 5G will be able to peak at 20 Gbit/s of speed. Additionally, some testing has already achieved rates of about 7 Gbit/s. Yet an actual average rate of more than 100 Mbit/s is anticipated. Even though some sources suggest that 4G can reach peak rates of hundreds of Mbit/s, there is still a significant gap between this and that technology.

B. **It is latency-efficient:** The time it requires for data to be transmitted between two sites in a network or between us issuing an order and when an event occurs has been significantly decreased with 5G. Unsurprisingly, it affects an estimated on how 5G functions. According to Ericsson research, most mmWave tests have demonstrated that 5G does have a latency of under 50 ms. However, certain analyses reveal that

4G often remains in the 30 to 50 ms range, whereas 5G has the potential to possibly reduce that figure by 10 times, to 3 to 5 ms.

C. **Greater simultaneous connectivity is made possible by its large bandwidth:** Due to the high bandwidth, it offers, 4G introduced the 800 MHz and 2600 MHz frequencies to boost data transfer rates. Then, 5G increased bandwidth by including the mmWave frequencies of 700 MHz and 3400 MHz to 3700 MHz in the RF (radio frequency) spectrum. Short-length guided telecommunication point-to-point connection used to be the only ones that this apparition could occur on.

Instead of having OFDM (orthogonal frequency-division multiplexing) encoding, which 4G lacked, 5G microcells up to 1 W in power with a short range, located every 100 to 300 metres, do. The cells cooperate, and hundreds of thousands or even millions of devices can really be integrated because with mobile networks and high bandwidth. For 5G, each nation may assign frequencies. The 5G frequency range in the European Union is 24.5 GHz to 27.5 GHz, though it may increase.

D. **Network slicing in 5G facilitates hyper-connectivity:** Network slicing, a characteristic of 5G microcells, enables them to categories equipment into 16 or more groups and assign each group the appropriate resources. For instance, the fastest speeds should be given to devices streaming entertainment or performing critical tasks. The speed of IOT gadgets, smart devices and sensors will be constrained. Network throttling can be resolved thanks to improved connectivity and higher capacity. However, 5G still guarantees each device's best performance.

E. **It has a wide range of applications:** 5G can take the place of many of the daily-use technologies due to its extraordinarily fast transmission speed. For illustration, the 5G technology of the whole broadband Internet network standard will do away with the requirement for a large of storage in PCs, TVs or portable devices because of how quickly it will switch. In the end, cloud storage offers quick data downloads. Additionally, unless such distance is too large, 5G's wireless application may replace optical cables because speed can be on par with or faster than that of optical cables. The same goes for installing software and buying PCs with processors, RAM and GPUs.

Edge and cloud computing via a terminal may become a more convenient, affordable and quick option with 5G. Ultra-accurate, ultra-rapid remote control is made feasible by the latency. For instance, with a strong 5G connectivity and lots of sensors, remote transportation systems like driverless cars could become a viable choice. That holds true for medical technology, virtual reality endeavours, multiplayer video games and even manufacturing robots. With the proper equipment, doctors may hypothetically do operations remotely. These programmers support remote learning and improved monitoring, boosting literacy and lowering crime. In a nutshell, 5G is a technology that raises standards of living.

2.5.4 Drawbacks of 5G

After learning about the advantages of 5G for the world, let's examine some drawbacks of the network [18]:

A. **5G's rollout is sluggish:** Although 5G has a lot of potential, the technologies are not yet widespread. While other nations must wait years before the technology even reaches them, some nations are implementing it more quickly. Additionally, implementation does not guarantee flawless performance; the new mobile network providers will encounter temporary technical issues. Additionally, the economic climate can dictate that only major cities get access to 5G. Rural areas could be put on wait once more for a few more years.

B. **The obstruction risk for its network towers is high:** The potential capabilities of 5G appear to be exceptional once more. While tall structures and trees obstruct the 5G network, those measurements were taken under ideal circumstances. Therefore, in places with towering objects, 5G may not be possible or may work poorly. Carriers are forced to obstruct rural regions from coverage by mounting the towers atop those items in order to combat this. If not, they will need to construct other towers nearby, which has drawbacks from both an economic and aesthetic standpoint.

C. **5G is not compatible with older devices:** The incompatibility of 5G with 2G, 3G and 4G devices is a major disadvantage. For people to benefit from 5G connectivity, they must replace equipment that are just a few years old. Because of this, the general populace cannot use technology. Additionally, it makes 4G gadgets instantly obsolete, producing more technological waste.

D. **5G has security issues:** Network and software components that comprehend data and allocate resources are necessary for network slicing. This opens the door for data espionage, which the US charged Huawei with carrying for the Chinese government around 2020 before banning them. Security researchers also discovered weaknesses that expose 5G networks to DDoS (distributed denial of service) and MITM (man-in-the-middle) attacks. It will be vulnerable to new exploits as a new technology. Additionally, malevolent users may make use of the extra bandwidth granted IOT devices, strengthening botnet attacks.

E. **5G could have detrimental consequences on health:** The health risks of 5G have received a lot of attention and have been repeatedly refuted. After all, merely a few metres away from the tower, the potentially hazardous effects soon diminish. The long-term impacts of 5G are unknown, particularly as the quantity of nearby cell towers increases. Without concrete evidence, concerns about how 5G can impact health shouldn't prevent the technology's adoption. However, it needs to be watched over and kept at levels that are suitable for populous regions.

2.6 5G APPLICATIONS

5G technology is embellished with numerous and distinctive qualities that are applicable to a wide variety of individuals regardless of their aims. Future industries powered by 5G technology include retail, education, transportation, entertainment and healthcare. The applications of 5G are shown in Figure 2.13 and discussed in the following sections [30].

A. **Autonomous Vehicles:** Rapid technological advancements in the automobile industry enable the development of automated cars. Due to the drastically reduced latency of 5G networks, which will allow for responses rapid than over existing cellular networks, autonomous vehicles will be greatly enabled. The goal is to create a vehicle-to-everything (V2X) communication network that will allow vehicles to nearly instantly react automatically to objects and differences in their vicinity. For a car to stop or change angle of driving in reaction to traffic signs, dangers and pedestrians crossing the road, it should be capable of transferring messages in milliseconds.

B. **Smart City:** Intelligent transportation systems are currently being implemented by several cities throughout the world, and linked car technology will be supported by these systems. Bidirectional communications between vehicles and infrastructure will be made possible by connected cars to make safety important throughout the vehicular system. In smart metropolitan settings, sensors are now being deployed in every intersection to detect changes in position and interconnect autonomous cars to respond as needed. Even before 5G is applied effectively, the communications infrastructure needed to support connected car innovation can be brought in, significantly enhancing both vehicle and pedestrian safeness.

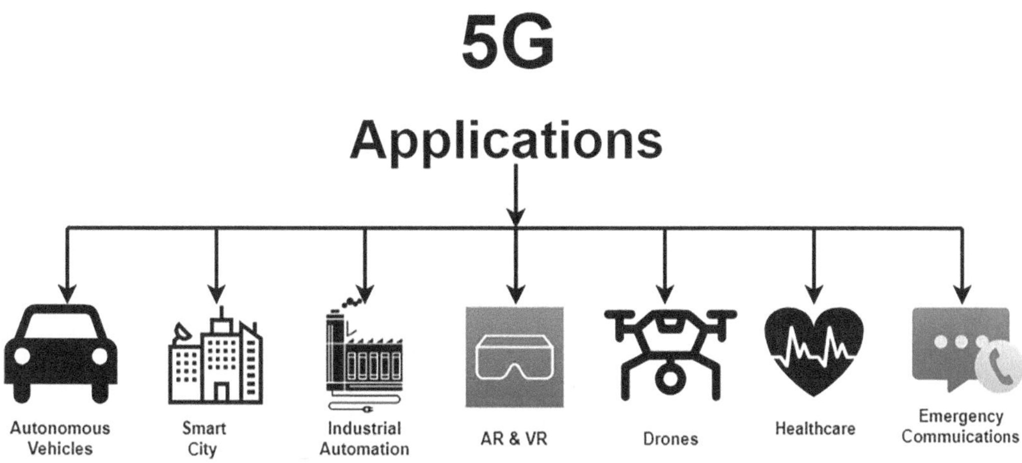

FIGURE 2.13 Applications of 5G technology.

C. **Industrial Automation:** Wireless adaptability, low cost and the reliability of networks that aren't possible with present wireless technology are the key points of 5G in the industrial autonomous sector. Smart manufacturing is currently in use, and you've probably seen movies of synchronized robotics in use in various production processes and sectors. Wi-Fi currently lacks the range, mobility and quality of service needed for industrial control, and current cellular technology has a high latency, making wires necessary for many applications. These applications can become entirely wireless with 5G, making it productive advance manufacturing.

D. **Augmented Reality (AR) and Virtual Reality (VR):** Applications for augmented and virtual reality will become much more realistic and engaging thanks to 5G's low latency. For instance, a mechanic using 5G AR eyeglasses in an industrial setting may view a footprint of a system that would detect parts, offer support guidance or highlight parts that shouldn't be handled. You can hold augmented reality meetings in professional settings where it looks that two individuals are seated side by side in the same space, converting dull phone or two-dimensional virtual meetings into more engaging three-dimensional events.

E. **Drones:** Beyond the consumer application for filmmaking and photographing, drones have a wide and expanding range of use cases today. Retailers and wholesalers are considering using drones to transfer products. The trend is going to continue, and 5G will enable us to push the current capabilities of drones, majorly in terms of distance and interactivity. A drone can take advantage of high-dependability and low-latency connectivity when using a 5G network. This indicates that a drone can quickly respond to instructions from the operator or ground control system.

F. **Healthcare:** The network will face difficult demands in terms of bandwidth, data rate and latency, among other things, as the medical sector has rapidly demanding market with a surge in a wide variety of applications that will use the wireless technology and use distinct types of data of different sizes and formats. Beginning with sensor devices in healthcare facilities that employ current technologies like Wi-Fi, Bluetooth and associated low-power technologies, massive-machine-type communication will expand as this market expands due to the connection requirements for the high demand of equipment and equipment with sensor-based uses in multi-specialty hospitals.

G. **Communications in an Emergency:** Lately, mobile edge computing has been suggested as a solution to fill this gap created by the existing cloud architectures. An efficient virtual environment that is located at the network's edge is made possible by MEC, which is regarded as a crucial enabling technology. The transition to 5G in telecommunications networks offers the chance to improve operations, support information exchange, increase efficiency and response time, bolster security and expand the capacity for more connected devices.

2.7 CONCLUSION

This study traces the development of wireless communication technology from its earliest iterations. The necessity of 5G as well as its difficulties and needs are also emphasized. It is desirable that research and development be done, anticipating future market trends, demands, societal issues, technological evolution and looking beyond the horizon of 5G, as 5G is projected to be utilized across many industrial domains. Although 5G has only recently begun to roll out in some areas and is still in its infancy, researchers and developers are already working on the development of 6G, particularly in relation to new communication methods and technologies. The upcoming mobile system, known as sixth generation, is anticipated to be introduced in 2030 and would fundamentally alter mobile wireless networks.

REFERENCES

1. Laghari, A. A., Riaz, M. M., & Shah, M. A. A. (2021). Analog Cellular Systems. In *Handbook of Mobile Communication Studies* (pp. 1–10). MIT Press. Doi: 10.7551/mitpress/12038.003.0004.
2. Qureshi, I. M., Qayyum, A., & Siddiqui, M. S. (2012). Evolution of 2G and 3G Wireless Networks. In *2012 International Conference on Advanced Computer Science Applications and Technologies* (pp. 450–455). Doi: 10.1109/ACSAT.2012.27.
3. Mellios, E., & Fountoulakis, D. (2017). Global System for Mobile Communications (GSM). In *International Encyclopedia of Public Health* (pp. 430–436). Elsevier. Doi: 10.1016/B978-0-12-803678-5.00062-4.
4. Dholakia, H. T., & Kosta, M. (2019). Analysis of GSM System Architecture and its Security Issues. In *2019 IEEE 5th International Conference on Computing Communication and Automation (ICCCA)* (pp. 1–5). IEEE.
5. Al-Janabi, S., & Ibraheem, A. (2019). An Overview of the GSM Network Architecture. In *2019 2nd Scientific Conference of College of Information Engineering (SCCIE)* (pp. 1–7). IEEE.
6. Chen, Y., & Perros, H. (2005). *Internet Telephony: Architecture and Protocols*. Springer Science & Business Media.
7. Gupta, A., & Garg, K. (2022). A Comprehensive Review of Fourth-Generation (4G) Technology. *Journal of Ambient Intelligence and Humanized Computing*, 13(3), 3445–3460. Doi: 10.1007/s12652-021-03412-0.
8. Prasad, R., & Muftic, S. (2012). 4G/LTE Mobile Communication Systems. In *Security for Mobility* (pp. 45–74). Springer.
9. Andrews, J. G., Buzzi, S., Choi, W., Hanly, S., Lozano, A., Soong, A. C. K., & Zhang, J. C. (2014). What Will 5G Be? *IEEE Journal on Selected Areas in Communications*, 32(6), 1065–1082. Doi: 10.1109/JSAC.2014.2328098
10. Mishra, A. R., & Rathore, S. S. (2019). 5G: Architecture, Components and Future. *2019 4th International Conference on Computing Methodologies and Communication (ICCMC)*, Erode, India, pp. 1247–1252. Doi: 10.1109/ICCMC.2019.8869158.

11. Ali, W., Saeed, A., Ur Rehman, M. H., Ali, I., & Khattak, A. M. (2019). A Comprehensive Survey on 5G: Key Enabling Technologies and Design Issues. *IEEE Communications Surveys & Tutorials*, 21(3), 2344–2390. Doi: 10.1109/COMST.2019.2913337.
12. Gao, X., Wang, X., & Yang, J. (2021). 5G and Beyond Wireless Networks: Challenges and Future Directions. *IEEE Access*, 9, 34722–34736. doi: 10.1109/access.2021.3060138.
13. Park, S., & Lee, J. (March 2017). Digital Beamforming in Multiuser MIMO for LTE Advanced Pro and 5G NR. *IEEE Communications Magazine*, 55(3), 178–184. Doi: 10.1109/MCOM.2017.1600839.
14. Jung, J., Park, J., & Lee, S. (2019). Beamforming for 5G Mobile Networks: Recent Trends and Future Research Directions. *Nature*, 9(1), 1–13. Doi: 10.1038/s41598-018-38304-2.
15. Huang, X., Li, S., & Li, J. (2018). Mobile Cloud Computing: Research Issues, Challenges, and Solutions. *Future Generation Computer Systems*, 81, 659–676. Doi: 10.1016/j.future.2017.11.036.
16. Zhang, Y., Wang, H., & Zhang, L. (2020). 5G and Beyond: Opportunities and Challenges. *Frontiers of Information Technology & Electronic Engineering*, 21(3), 373–385. Doi: 10.1631/FITEE.1900685.
17. Bhat, Aparna, Gojanur, Vishwanath, & Hegde, Rajeshwari. (2015). *5G Evolution and Need: A Study* (pp. 1–4). Doi: 10.1109/EESCO.2015.7253944.
18. ITU. (2019). Report on Spectrum for 5G. https://www.itu.int/en/ITU-D/ICT-Applications/Documents/Spectrum_for_5G.pdf.
19. Uusitalo, M., Rugeland, P., Boldi, M. et al. (2021). 6G Vision, Value, Use Cases and Technologies from European 6G Flagship Project Hexa-X. *IEEE Access*, 9, 160004–160020. doi: 10.1109/ACCESS.2021.3130030.
20. Gawas, A. U. (May 2015). An Overview on Evolution of Mobile Wireless Communication Networks: 1G-6G. *International Journal on Recent and Innovation Trends in Computing and Communication*, 3(5), 3130. Doi: 10.17762/ijritcc.v3i5.4404.
21. Mourad, A., Yang, R., Lehne, P. H., & De La Oliva, A. (2020). A Baseline Roadmap for Advanced Wireless Research Beyond 5G. *Electronics*, 9(2), 351. doi: 10.3390/electronics9020351.
22. Solyman, Ahmed, & Yahya, Khalid. (2021). Key Performance Requirement of Future Next Wireless Networks (6G). *Bulletin of Electrical Engineering and Informatics*, 10, 3249–3255. Doi: 10.11591/eei.v10i6.3176.
23. Zhang, H., & Mao, S. (2013). *LTE-Advanced: A Practical Systems Approach to Understanding 3GPP LTE Releases 10 and 11 Radio Access Technologies*. Academic Press.
24. Asghar, M. Z., Memon, S. A., & Hämäläinen, J. (2022). Evolution of Wireless Communication to 6G: Potential Applications and Research Directions. *Sustainability*, 14(10), 6356. doi: 10.3390/su14106356.
25. Rasti, M., Taskou, S. K., Tabassum, H., & Hossain, E. (2021). Evolution Toward 6G Wireless Networks: A Resource Management Perspective. *ArXiv, abs/2108.06527.* https://scholar.google.com/scholar?hl=en&as_sdt=0%2C5&q=Rasti%2C+M.%2C+Taskou%2C+S.+K.%2C+Tabassum%2C+H.%2C+%26+Hossain%2C+E.+%282021%29.+Evolution+Toward+6G+Wireless+Networks%3A+A+Resource+Management+Perspective.+ArXiv%2C+abs%2F2108.06527.&btnG=#d=gs_cit&t=1712212004866&u=%2Fscholar%3Fq%3Dinfo%3A9YQMojNIHqEJ%3Ascholar.google.com%2F%26output%3Dcite%26scirp%3D0%26hl%3Den.
26. Mozaffariahrar, E., Theoleyre, F., Menth, M. (2022). A Survey of Wi-Fi 6: Technologies, Advances, and Challenges. *Future Internet*, 14(10), 293. doi: 10.3390/fi14100293.
27. Pujari, V., Patil, R., & Tambe, K. (2021). Research Paper on Future of 5G Wireless System. *Contemporary Research in India*. https://scholar.google.com/scholar?&q=Pujari%20V%2C%20Patil%20R%2C%20Tambe%20K%20%282021%29%20Research%20paper%20on%20future%20of%205G%20wireless%20system.
28. Dangi, R., Lalwani, P., Choudhary, G., You, I., & Pau, G. (2022). Study and Investigation on 5G Technology: A Systematic Review. *Sensors*, 22(1), 26. doi: 10.3390/s22010026.

29. Alraih, S., Shayea, I., Behjati, M., Nordin, R., Abdullah, N. F., Abu-Samah, A., & Nandi, D. (2022). Revolution or Evolution? Technical Requirements and Considerations towards 6G Mobile Communications. *Sensors*, 22(3), 762. doi: 10.3390/s22030762.
30. Osseiran, A., Monserrat, J. F., & Marsch, P. (2016). 5G Mobile Networks: Vision, Requirements and Challenges. *IEEE Communications Magazine*, 54(5), 1–9. Doi: 10.1109/MCOM.2016.7470937.

CHAPTER 3

Metaheuristic Algorithms and Its Application in Enterprise Data

Radhika D. Joshi, Sheetal Waghchaware and Rushikesh Dudhani

3.1 INTRODUCTION

This chapter showcases a variety of enterprise data generated by different applications like healthcare, business, industry [1], agriculture [2] and security. It is essential to explore the techniques required to extract the information from this big data. These applications demand the collective representation and comparison of different metaheuristic algorithms for solving real-world complex problems. It catches the attention of researchers working in the enterprise data domain. Some of the popular metaheuristic algorithms are genetic algorithm (GA), ant colony optimization (ACO) [3], particle swarm optimization (PSO), evolutionary algorithms (EA) and tabu search (TS). Most of the metaheuristic algorithms are nature-inspired, while few of them are trajectory-based and evolutionary based. Simulated annealing is trajectory-based metaheuristic algorithm published in year 1983, while genetic algorithm published in year 1992. The progress of metaheuristic algorithm and its variants is started by this. Fire hawk optimizer [4] is metaheuristic algorithm recently published in year 2022. To date hundreds of such metaheuristic algorithms are available. In order to decide the usability of these algorithms, system performance measurement is must. A set of performance evaluation metrics required for analysis are specified and algorithms are tested for total accuracy parameter. Dataset plays vital role in order to examine the performance of system. Public datasets available for metaheuristic are travelling salesman problem dataset, knapsack problem dataset, quadratic assignment problem dataset and capacitated vehicle routing problem dataset. UCI (University of California, Irvine) maintains several dataset repositories that are widely used in the research community for testing, benchmarking and developing new machine learning algorithms, which contains more than 500 datasets [5]. This chapter is organized using following main points role of enterprise data, need and types metaheuristic algorithms, use of computer vision (CV) for enterprise applications,

DOI: 10.1201/9781032699806-3

and hyperparameters applicability in metaheuristic optimization. Experimentation is carried out on a set of three algorithms and ten different datasets for the result analysis.

3.2 ROLE OF ENTERPRISE DATA

In the present time, data is said to be the foundation for business to be successful. Business data can be used to learn more about customer preferences, grievances and company operations. It provides real-time intelligence that facilitates to generate more accurate market projections for the industry.

Enterprise data management is essential as it promotes decision-making based on data analysis consecutively aiding to run business operations more successfully. Enterprise data management facilitates better data quality in terms of accuracy and reliability of the information. It enables better business intelligence and enhances efficient operational intelligence. It also helps to reduce the costs of IT infrastructure, storage and backup. In order to accomplish these tasks, it is essential for the enterprises to extract meaningful insights from unstructured data. Unstructured data is the information that does not follow a data model and has no easily identifiable structure and therefore cannot be used in the same format. In the present time, enterprises are switching over to AI so as to unblock more business value and enhance their operations. AI tools are essential to assist businesses to effectively manage the unstructured data and gain value from the same; e.g., pattern recognition algorithms perform the work to categorize unstructured data. Thus, metaheuristic optimization algorithms help to provide best solution.

Figure 3.1 shows the block diagram of AI-based model for handling enterprise data. The enterprise data is a big data that requires to be stored in various data storage devices such as network-attached storage, direct-attached storage, cloud storage, etc. The raw

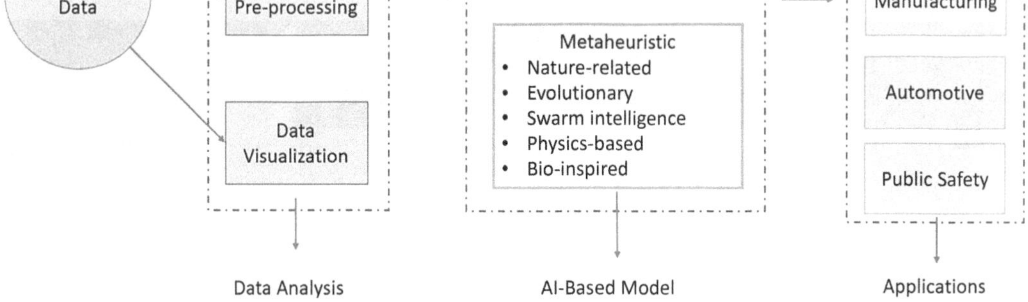

FIGURE 3.1 Architecture of AI-based model for enterprise data.

enterprise data has to be pre-processed by using various methods, such as data cleaning, data transformation and handling imbalance data, followed by dimensionality reduction methods. After the pre-processing of data, it is visualized with help of data visualization techniques that includes bubble plot, scatter plot, histogram, pie chart, spider chart, bar chart, waterfall chart, table, etc. The different software tools available for big data analytics includes Accenture, Alpine Data, Alteryx, Angoss, BigML, BIME, Clario and CoolData [5]. Appropriate machine learning (ML) or deep learning (DL) algorithms are applied to train the model by using various computer vision tools like OpenCV, Viso Suite, SimpleCV and BoofCV. Optimization is a technique for delivering best solutions in order to increase the accuracy and reduce the loss of model. At the core of nearly all ML and DL algorithms, there is an optimization algorithm. The training efficiency of the model gets influenced by optimization techniques and it helps to achieve the best possible output. Metaheuristic algorithms are used for optimization of model and it is helpful for solving real-world problems. Metaheuristic algorithms are classified as evolutionary, swarm-intelligence, physics-based and bio-inspired algorithms. It can also be implemented in numerous applications like healthcare, manufacturing, automotive, public safety, media and entertainment.

3.3 STEPS FOR CONVERTING DATA INTO INFORMATION

Extracting meaningful and useful insights is essential to facilitate better business decisions. The steps for extracting insightful information from the data is shown in Figure 3.2

1. **Compiling all raw big data:** The compilation of organized, unstructured and semi-structured data collected from any modality related with healthcare, biomedical, finance, retail, cyber security, networking, supply chain management, manufacturing, etc. is known as big data. Big data is practiced by businesses to enhance internal business operations. Per standards, big data are represented by volume, velocity and value wherein, volume represents the size of the data, velocity provides the speed of data and the value indicates its significance to enable the process of taking actionable decisions.

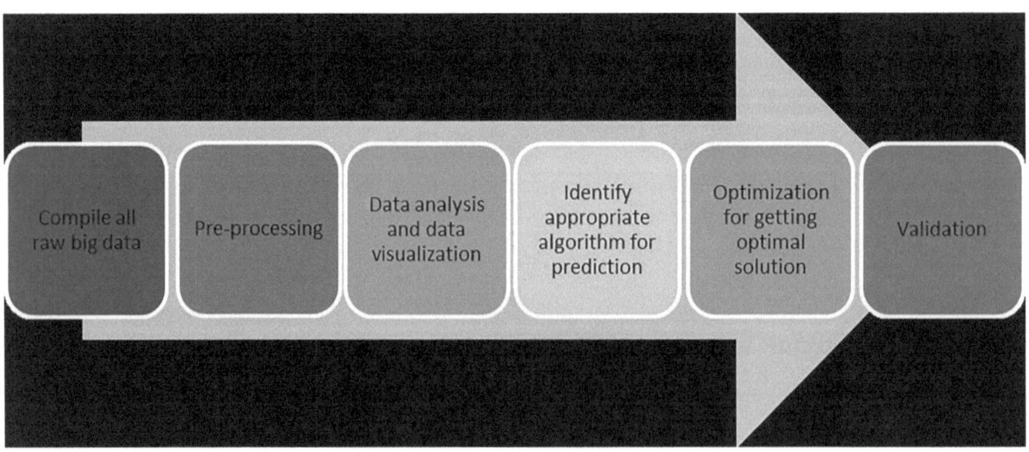

FIGURE 3.2 Extraction of meaningful information from data.

Vast amount of data has to be gathered so as to identify the patterns present in the data by making use of direct and indirect sources to improve the business [6]. Operations data for supply chain management, manufacturing, marketing plans, analysis of employee behaviour, etc. is generated with the help of direct data sources in enterprises. A few examples of indirect information are data gathered through click streams, geo-location data, health records and many more categories of data which are in relation with the customer. The third-party suppliers for data are also available such as database marketers or market research companies. Businesses generally make use of big data for market research, customer segmentation and personalization. Similarly, enterprises gather information from e-commerce websites to examine client reviews and ratings.

2. **Pre-processing and data cleaning:** Pre-processing and cleaning of data is considered as mandatory before redirecting the same for data analysis. Data cleaning is considered very essential as it can put forward misleading or incorrect insights. It can be potentially devastating if these insights are considered to make crucial business decisions. Data-cleaning techniques that need to be followed in sequential manner are given in Figure 3.3. It starts with preliminary stage of clear formatting, consecutively goes through various stages and reaches up to the last stage- convert data type.

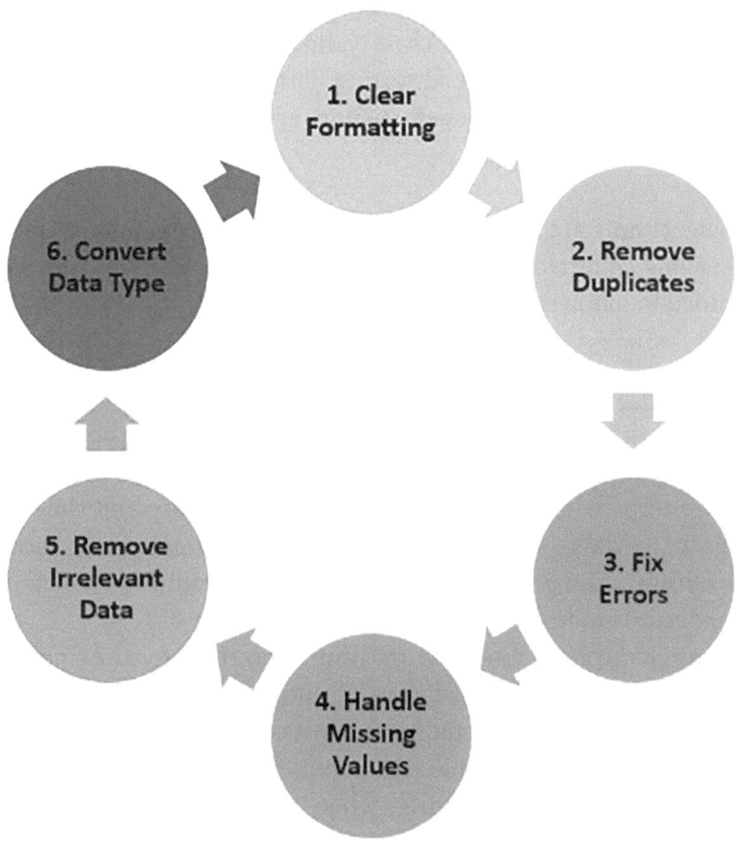

FIGURE 3.3 Data-cleaning techniques.

3. **Data analysis and data visualization:** Data analysis helps in extracting useful information from the big data. It assists to identify the target customers, saves time, reduces operational costs and generates more accurate data. The various tools accessible for data analysis includes Excel, Python, R, Looker, Rapid Miner, Metabase, Redash and Microsoft Power BI.

   The selection of the appropriate visualization method for data insights is considered equally important. Choosing the best suitable techniques completely relies upon the type of visualization requirement to access between various datasets and is decided upon the relationship, distribution, composition and comparison.

4. **Identify appropriate algorithm for prediction:** Predictive analysis makes use of computer vision (CV) and its various algorithms. Different ML- and DL-based CV techniques provides better performance on the enterprise data for prediction.

5. **Optimization for getting optimal solution:** The process of optimization improves the accuracy of system and by default also reduces errors. AI-based metaheuristic optimization algorithms facilitate optimal solutions which provide the best outcome to the problem. The optimal solution provides profit maximization, cost minimization and accurate prediction.

6. **Validation:** The resultant validation can be processed with the assistance of various validation methods such as cross-validation, fold validation and separating the dataset into training data and testing data. The outcome can be validated by assessing performance parameters such as accuracy, F1-score, precision, recall, selectivity, etc.

Metaheuristic algorithms are needed for the analysis and deduction of conclusions from the valuable insights collected from massive data. The necessity of metaheuristic algorithms is explained in following section.

3.4 NEED OF METAHEURISTIC ALGORITHM

A metaheuristic algorithm proves to be a good solution to complex optimization problem, while it is a challenging situation to solve the same by other search procedures. In the world of limited means with reference to processing capability and time, metaheuristic algorithm proposes flawless solutions which are founded on imperfect or incomplete information. While optimization problems extend in scale and complexity, metaheuristics will prove to be more universal in the coming days.

Concisely, though the comprehensive optimum solutions for the final metaheuristic cannot be assured, it may be at least worthy enough to be trusted for the approximated solutions within computational time limits. Metaheuristics facilitates answers to inevitably unsolvable, unattainable and complex problems. Metaheuristic delivers approximation of the optimal matching, when attaining exact solutions is difficult to accomplish within rational computation time. Time and scalability are important factors to be studied in systems where a radical issue like quality of output is to be reviewed.

Metaheuristic basically wheels around extending flexibility to the ways of utilizing control parameters so that it can be appropriate to the nature of the problem under consideration. This is achievable by encompassing highly scalable, reliable intelligent techniques and practices of problem-independent algorithmic frameworks to resolve complex working situations for the end users.

Nature-inspired optimization algorithms can cover multidisciplinary, widely varied fields of problems and objectives. These algorithms can be applied efficiently than the conventional optimization methods.

3.5 TYPES OF METAHEURISTIC ALGORITHMS

Fred W. Glover was the one, who first coined the term "metaheuristic" in his research study [7]. Metaheuristic search methods are higher-level general approaches that can be introduced as guiding principles when designing the basic heuristics to acknowledge particular optimization concerns. Metaheuristic algorithms is included in the approximation optimize techniques which is most popular techniques for the past three decades [8] and is used to resolve big and higher complexity issues.

Metaheuristic algorithm is mainly classified under ten groups as shown in Figure 3.4. Several metaheuristics draw their inspiration from nature, bio and physics [9]. A few metaheuristic algorithms do not require memory, which suggests they do not explore the information that is assembled dynamically. Local search, GRASP and simulated annealing are some of the examples of this class. On the other hand, some metaheuristics make use of memory that stores data, collected from the Internet during the search.

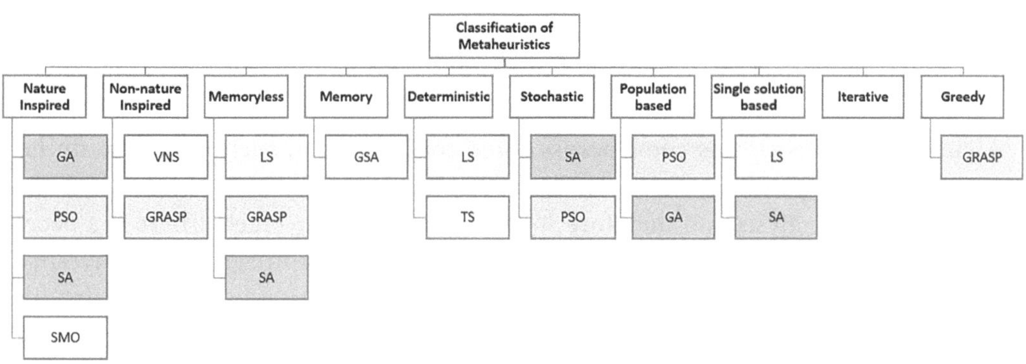

FIGURE 3.4 Classification of metaheuristic algorithms. GA: genetic algorithm, PSO: particle swarm optimization, SA: simulated annealing, SMO: social mimic optimization, VNS: variable neighbourhood search, GRASP: greedy randomized adaptive search procedure, LS: local search, GSA: gravitational search algorithm, TS: tabu search

Deterministic actions, with the help of deterministic metaheuristics, are used to resolve optimization problem. Random rules are applied by stochastic metaheuristics that permit them to explore many answers from the same initial solution. In case of deterministic algorithms, the same initial solution will result in the same end solution. This feature has to take into consideration for investigating the effectiveness of metaheuristic algorithms [10]. Single-solution based metaheuristics stress on exploitation, while population-based metaheuristics focus on exploration. These two groups experience complementing characteristics [10]. The iterative algorithms begin with a complete solution and then modify it at each repetition. Greedy algorithms commence with an incomplete solution and attach a decision variable to the issue at each stage until they encounter a complete solution.

The literature accessible shows that the integration of metaheuristic algorithms and deep learning has provided advantage in enhancing the system performance. In addition, with the implementation of the optimization algorithm, the parameters needed for deep learning model can be minimized. This ultimately results in lowering the human efforts.

The metaheuristic algorithms can be organized as evolutionary, bio-related, nature-related, swarm-intelligence, physical or mixed-approach. Most of the metaheuristic algorithms are biologically influenced algorithms, which are structured from the behaviour of evolutionary biology of living things [5]. Bio-related algorithms stress more on distributed, decentralized, self-organized, flexible intelligence evolved from the nature algorithms. The nature-related algorithms may possibly pull-out inspiration from the biological system, evolution, human activities, behaviour of animal groups, etc. They can address highly non-linear and complicated issues, specifically in the streams of engineering and science. They are presented to be very effective and efficient in managing real-world optimization problems. Swarm intelligence algorithms gather their ideas from the social interactions of birds, insects and other living creatures. The phenomena of the natural world, such as Newton's law of gravity and the heating and cooling of substances, provide inspiration for the physics-based methods. When these are implemented to the challenges in the real world, these physical algorithms have delivered encouraging results. The metaheuristic algorithms with its revelation, implementation areas and its challenges are summarized in Table 3.1.

Metaheuristic algorithms come across some confrontation, such as lack of unified framework, the expanding metaheuristic applications are more advanced than that of mathematical analysis and therefore solutions can't be converged. There is a necessity to categorize the algorithms as well as need of unified approach for the investigation. Metaheuristic optimization algorithms takes its inspiration from the nature. The noticeable types are (a) bio-inspired (they make use of biological evolution), (b) science-inspired (basically, they are influenced by physics and chemistry), (c) art-inspired (they are useful for the global optimization and are influenced from the artists conduct and actions) and lastly (d) socially influenced (they encourage social behavior to resolve optimization) [4].

TABLE 3.1 Summary of Different Metaheuristic Algorithms

Category	Algorithm	Principle of Operation	Application Domain	Limitations
Bio-related [11–13]	**Bacterial Forging**	The way *Escherichia coli* bacteria forage	Mathematical analysis, pattern recognition	Fixed step size, weak connection among bacteria leads to local optimum
	Genetic Algorithm	Inspired from natural evolution and genetics	Image processing, vehicle routing problems, parametric design of aircraft	Expensive, lack of memory, not effective for simple problems
	Social Mimic Optimization	Influenced from social behaviour mimicry	Mathematical computations, data computations	*
	Dendritic Cell	Human immunity structure, dendritic cell function	Robotic algorithm, dendritic cell population management	*
	Krill Herd	Antarctic krill are on a move to look for food and gather more of it.	High accuracy compared to other optimization techniques	Poor local search
	Artificial Humming Bird	Behaviour of specific flight skills of hummingbirds	Shows superior effectiveness in terms of computational burden and solution precision compared with existing techniques	Unable to tackle complex multi-objective optimization problems and engineering design problems
	Corona Virus Optimization	Behaviour of novel COVID-19	To recognize the various ways, it may impact on the health of people	Difficult to implement
	Grey Wolf Optimization [5]	Hunting process involving pack of grey wolves	Valuable for searching optimized solutions and for classification task	Lack of population diversity, poor exploration capability
	Strawberry [5]	Plant movement governed by strawberry plant reproduction	Simultaneously investigates multiple regions of the exploration area and enables to depart from the local maximum to reach the global optimal state	Precision depends on initial parameters and maximum iteration number.
	Spider Monkey	Intelligent foraging behaviour of spider monkeys	Balances between exploitation and exploration while search for the optima	Difficulty in obtaining an optimal solution for classification problems

(*Continued*)

TABLE 3.1 (Continued)

Category	Algorithm	Principle of Operation	Application Domain	Limitations
Nature-Related [14, 15]	**Cuckoo Search**	Behaviour of cuckoo birds and their reproduction	Used to generate new candidate solution that solve optimization problems	The weak local search capability contributes to the sluggish convergence.
	Flower Pollination [6]	Pollination procedure of flowers	Developed to solve feature selection problem	Tendency towards premature convergence and poor exploitation ability
	Invasive Weed	Behaviour of weed colonies	Approximate in terms of global search, optimization problems – antenna design, telecommunication design, DC motor control	Premature convergence when solving complex optimization problems
	Manta Ray Foraging	Imitates manta rays' feeding habits	Stress design, beam design, speed reducer design, data computations	Deficiency of decreasing population diversity, weakens exploitation capability
	Sailfish Optimizer	Repetitive attack method used by sailfish during team hunting	Solves engineering optimization problems	Accuracy is not sufficient; convergence speed is less.
	Bat Algorithm [6]	Echolocation nature of bats	Approved for resolving several real-world issues	High probability of being trapped in local optima due to poor exploration
Swarm Intelligence [16]	**Ant Colony Optimization**	Ants' ability to pick out the quickest path amidst food and their anthills.	Best for combinational optimization and addressing graph problems	Stagnation, lengthy search time, local optimum problem
	Particle Swarm Optimization	Concept of swarm intelligence	Used to solve complex mathematical problems	Easy to fall into local optimum in high-dimensional space, parameter selection
	Artificial Bee Colony	Forging behaviour of honey bee	Robot control optimization, routing problem	Insufficient population diversity, weak developing capacity
	Fish Swarm	Normal, united motion of fish, and their social habit.	Fuzzy logic controller, image processing, Scheduling	Imbalance among global and local search, inadequacy to gain from the events of peers for the upcoming action

Physical [17, 18]	**Shuffled Frog-Leaping**	Natural metrics-inspired cooperative search metaphor	To solve a problem using combinatorial optimization	Easily fall into the local optimal solution and premature convergence
	Memetic	Population-based hybrid genetic algorithms	Ability to explore the solution space thoroughly while avoiding exhaustive search	*
	Simulated Annealing	Behaviour of systems with a lot of freedom when they're in a state of stability	In multi-component or connective gain	Each time this algorithm only works on a point so it cannot provide a general form of the search and it is without memory.
	Harmony Search	Process that the musicians use to improvise a perfect state of harmony	Salesperson navigation issues, educational development issue	Inability to provide an effective equilibrium among universal and regional exploration
Mixed-Approach [19, 20]	**Firefly Algorithm**	Process of firefly reproduction and commutation of message with the help of flashlights	Gives quick solution to problem	High computational time complexity, slow convergence speed
	River Formation	Copying how water forms river by eroding ground and depositing sediments	Telecommunications, software testing, industrial manufacturing processes, or navigation	*
	Cat Swarm Optimization	Resting and tracing behaviour of cats	Novel technique in modelling exploration and exploitation phases	Premature convergence problem

The summary of metaheuristic algorithms described in Table 3.1 shows that some algorithms have limitations, marked with an asterisk (*). This indicates no specific limitation can be highlighted and hence not mentioned.

Metaheuristic algorithms show resemblance in their structures. They can be categorized into two main group: evolutionary algorithms and swarm algorithms.

Swarm intelligence-based algorithms make use of the behaviour of animals. The well-known algorithms in this category are particle swarm optimization (PSO), ant colony optimization, bee algorithms, firefly algorithm, cuckoo search, bat algorithm, charged system search and krill herd.

Metaheuristic algorithms are used to optimize various computer vision (CV) tasks, including feature selection, object detection, image segmentation and neural network architecture optimization. In turn, computer vision techniques can be used to pre-process data for metaheuristic algorithms to improve their performance.

3.6 COMPUTER VISION FOR ENTERPRISE APPLICATIONS

CV is a subdivision of artificial intelligence that understands the way computers "see and understand" digital visual media. During the last decades, visual media and formats have encountered exceptional development. Corporations that require solutions can acquire customized solutions enhanced with complete security and privacy assembled into it. Computer vision systems have initiated the use in various industries like manufacturing, retail, healthcare, insurance, automotive, agriculture and public safety. These applications generate huge amount of data which need to be analysed in properly for extracting useful information with the help of metaheuristic algorithms.

Computer vision assists to process, evaluate and respond to digital images and video using machine learning (ML) or advanced analytics. The pixel data seized by the cameras is converted into matrix of numerical values in computer vision. This data is processed at the edge or sent to the cloud, where trained algorithms or software techniques analyse and identify the patterns from the image data. Large numbers of sample images are used for training through the statistical learning algorithms and neural networks. It is practicable to automatically detect and classify image patterns. Depending upon condition detected, it either alerts or initiates follow-up actions. Computer vision enables to provide thorough understanding and complete information even in case of complex situations like partially obstructed views, 3D modelling-based views, extrapolating data under unseen or hidden conditions.

In comparison, artificial intelligence takes a step forward to exceed in capabilities with human vision related to detection and processing. The presence of novel momentum in digital transformation accepted by the organizations during the last five to six years has witnessed new paths to upgrade performance and enhance welfare for the human race. Organizations opting for computer vision methods for working solutions can improve the process and the employee and customer experience, which in turn can be also assisting in increasing revenue. Different CV tools are available, such as OpenCV, SimpleCV, Tensorflow, MATLAB®, CUDA and BoofCV [21].

CV systems can be capable to execute several core functions, which delivers solutions to many challenging situations. The necessary potentiality encompasses image classification, localization, object detection, object counting, object tracking, image segmentation and optical character recognition. This drives the usage of CV in absolutely any sphere like energy and resources, manufacturing, retail and healthcare [22].

Metaheuristic algorithms often provide solutions near to correct optimal solution. This part is very important for optimizing the output of machine learning (ML) model. Deep learning (DL) method is profoundly useful for computer vision application situations as it can work effectively for recognition, visual tracking, semantic segmentation and image restoration. Since 2014, the eight emerging deep learning architectures – Alex Net, VGGNet, Google Net & Inception, ResNet, Dense Net, Mobile Nets, Efficient Net and RegNet—are most popular. Since 2019, the study trend signifies that CV situations are modifying from supervised learning to semi-supervised learning, which requires thorough knowledge of network type and architecture. Application scenarios are becoming more specific and demand refinement of techniques and cues compared to conventional CV. With the combination of CV with ML domain, it is possible to recognize communication from complex expressions. Enrichment of crossover applications expands the work scope in medical, industrial and forecasting prediction. Visualization of model and interoperability of DL models assist in managing large sizeable dataset by CV technique, which will be otherwise difficult. The scalability feature in the developed model desire to be considered from the early stage of design as in the later stages, it may lead to time consuming process during the training and testing stages [23, 24].

ML or DL models perform stepwise approach such as pre-processing, feature extraction, feature selection, optimization, model training, validation on unseen database. The efficiency of model is evaluated with the help of performance measures like accuracy, F-1 score, precision, recall, AUC value and specificity.

3.7 PERFORMANCE EVALUATION METRICS

The performance of a model can be examined with the help of performance metrics and these metrics are generalized for all algorithms. The various evaluation metrics required to analyse the model are listed in the following.

To resolve the classifications as well as regression challenge, confusion matrix is most familiar and frequently used. It is also known as error matrix or matching matrix and provides a visualization of the performance of an algorithm. It gives a summary of all classes that are correctly classified as well as incorrectly classified.

True positive (TP) value predicts positive from the observations of positive class.

True negative (TN) value predicts negative from the observations of negative class.

False positive (FP) value means for which model wrongly calculates observations in the positive class.

False negative (FN) value gives for which model wrongly calculates observations in the negative class.

1. **Accuracy**: Accuracy is the correct predictions, measured over the total prediction

$$Accuracy = \frac{TP + TN}{TP + TN + FP + FN} \quad [3.1]$$

2. **F-1 score**: The harmonic mean of recall and precision is calculated as the F-1 score.

$$F-1\,Score = \frac{Precision * Recall}{Precision + Recall} \quad [3.2]$$

3. **Precision**: Precision generally measures the degree of effectiveness of the model in categorizing the samples as positive by applying a positive predictive value.

$$Precision = \frac{TP}{TP + FP} \quad [3.3]$$

4. **Recall/Sensitivity**: It calculates the model's potential to locate positive samples.

$$Recall = \frac{TP}{TP + FN} \quad [3.4]$$

5. **Specificity**: It measures the quantity of appropriately identified negative occurrences to all true negative occurrences.

$$Specificity = \frac{TN}{TN + FP} \quad [3.5]$$

6. **AUC (Area Under Curve)**: It represents the probability of the model differentiating study from the two classes. It gives the probability of separation measured between classes. The value falls between 0 and 1. For example, in a multi-class problem, plot N number of curves for N number of classes.

$$AUC = \int_0^1 TPR(FPR^{-1}(x))dx \quad [3.6]$$

The evaluation metrics discussed in this section are responsible for deciding the performance of model. This performance of model has significant impact of hyperparameters, and optimizing them is a key part of the model tuning process.

3.8 HYPERPARAMETERS IN METAHEURISTIC OPTIMIZATION

In metaheuristic optimization, hyperparameters are parameters that manage the behaviour and performance of the metaheuristic algorithm. They determine the settings for the algorithm, such as the population size, mutation rate, crossover rate, number of generations, and stopping criteria [25, 26]. These hyperparameters hold a significant impact on the efficiency and effectiveness of the metaheuristic optimization, as well as locating the appropriate values for them is essential for attaining favourable outcomes.

In metaheuristic optimization, hyperparameters may differ depending on the specific algorithm, but some common ones consist of the following:

1. **Population size**: The sum of candidate solutions in the population.
2. **Mutation rate:** The possibility of a candidate solution undergoing a mutation operation.
3. **Crossover rate**: The possibility of two applicant solutions undergoing a crossover operation to produce offspring.
4. **Number of generations**: The highest number of iterations or generations the algorithm will run for.
5. **Stopping criteria**: The circumstances under which the algorithm will halt, such as when reached a certain fitness value or reached a maximum number of generations.
6. **Initialization method**: The method practiced to generate the initial population of candidate's solutions.
7. **Cooling schedule**: The schedule applied to control the cooling rate in simulated annealing.
8. **Pheromone trail initialization**: The technique practiced to initialize the pheromone trails in ant colony optimization.
9. **Exploration versus exploitation trade-off**: The equilibrium between examining new solutions and exploiting existing good solutions in particle swarm optimization.

These are some of the most common hyperparameters in metaheuristic optimization, and the accurate set of hyperparameters will rest on the specific algorithm being used.

Metaheuristic algorithms explore the space of possible hyperparameters and find the combination that yields the best performance for the given problem. This process is often referred to as hyperparameter optimization, or tuning.

3.9 EXPERIMENTATION

In metaheuristic optimization, the role of the hyperparameters is to control the behaviour of the optimization algorithm, while the confusion matrix carries out the evaluation of the results from the algorithm.

A confusion matrix is in a form of a table utilized to examine the performance of a classification model by comparing the actual values and the predicted values. The matrix is used to measure various metrics like accuracy, precision, recall and F1-score that are beneficial in assessing the performance of a model.

In connection with the metaheuristic optimization, the hyperparameters are utilized to manage the behaviour of the optimization algorithm and the objective function is to endeavour to optimize. The outcomes of the optimization can then be analysed with the help of confusion matrix, wherein the possibilities made by the optimized model are compared with the real numbers to determine its performance. In this context, the confusion matrix and hyperparameters in metaheuristic optimization are in relation, where the hyperparameters play a vital role in determining the interpretation of the optimized model, which can be further examined with the help of confusion matrix.

Figure 3.5 shows confusion matrix of hyperparameters for selecting set of best parameters. One hot encoding technique is applied for pre-processing data. The classification is performed by using multi-layer perceptron (MLP) classifier and best parameters are found with help of grid-search cross-validation technique. The dataset is split into train and test data. The training accuracy obtained is 88.16%, while test accuracy obtained is 87.61%.

The interpretation of three algorithms – PSO, GA and SMO – were validated with regards to the total accuracy. The ten datasets chosen for the experimentation were Breast Cancer (Dt1), Waveform (Dt2), Congress (Dt3), Exactly (Dt4), EWHeart (Dt5), Ionosphere EW (Dt6), Lymphography EW (Dt7), Sonar (Dt8), Spect EW (Dt9) and Waveform EW (Dt10). They are selected from UCI public dataset as they are open source and are already pre-processed. All these datasets are with reference to the medical domain and detailed information regarding them is mentioned in Table 3.2. They were utilized for the trials of algorithms. Two algorithms from the bio-inspired category – GA being the most popular and SMO being the least popular – were selected, while the other PSO algorithm from the stochastic optimization category was used for the comparison. The Jupyter Notebook platform was used as it facilitates better visualization of pre-developed code.

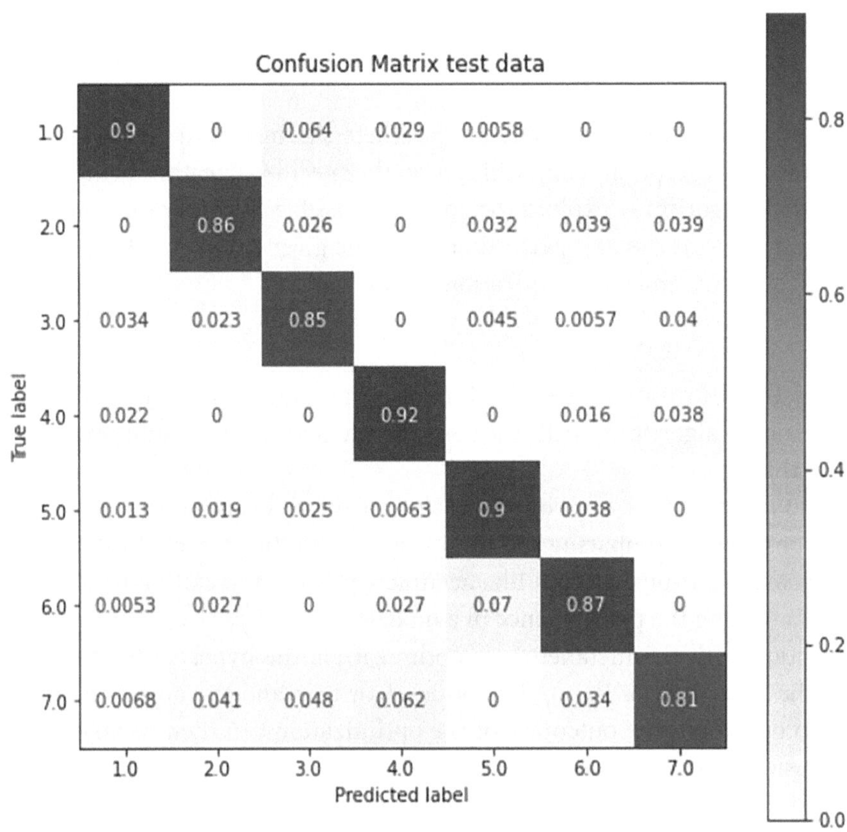

FIGURE 3.5 Confusion matrix.

TABLE 3.2 Dataset Information

Sr. No	Name of Dataset	Data Type	Attribute Type	Instances	Attributes	Missing Values
1	Breast Cancer (Dt1)	Multivariate	Categorical	699	11	Yes
2	Waveform (Dt2)	Multivariate, Data-Generator	Real	5000	21	No
3	Congress (Dt3)	Multivariate	Categorical	435	16	Yes
4	Exactly (Dt4)	Univariate	–	1000	–	–
5	EWHeart (Dt5)	Multivariate	Categorical, integer, real	303	14	yes
6	IonosphereEW (Dt6)	Multivariate	Integer, real	351	34	No
7	LymphographyEW (DT7)	Multivariate	Categorical	148	18	No
8	Sonar (Dt8)	Multivariate	Real	208	60	–
9	SpectEW (Dt9)	Multivariate	Categorical	267	22	No
10	WaveformEW (Dt10)	Multivariate	Categorical	1234	22	Yes

The total accuracy parameter for the three algorithms is illustrated in Table 3.2. Total accuracy is the accuracy calculated for only one dataset whereas average accuracy is calculated over ten datasets to determine the most suitable metaheuristic algorithm towards accuracy parameter. However, it is important to keep in mind that the accuracy of an algorithm can differ greatly, depending on the problem being optimized and the specific implementation used. This dataset provides a snapshot of the performance of the algorithms on a specific set of problems, whereas more comprehensive and controlled experiments would be required to draw broader conclusions about their relative performance.

With reference to the available literature, it is noticed that PSO is the most popular algorithm used, and this evidence is verified per the results obtained from Table 3.3.

TABLE 3.3 Comparison of Algorithms

	Total Accuracy		
Datasets	**PSO**	**GA**	**SMO**
Dt1	96.43	99.21	97.86
Dt2	94.74	94.74	95.61
Dt3	98.85	96.55	98.85
Dt4	76	85	76.00
Dt5	85.19	85.19	77.78
Dt6	97.14	97.14	91.43
Dt7	90.00	93.33	86.67
Dt8	92.86	95.24	90.48
Dt9	88.89	94.44	88.89
Dt10	81.10	90.43	79.70
Average Accuracy	**91.85**	**91.49**	**89.18**

3.10 SUMMARY

Enterprise data operates in a complex and dynamic environment. Metaheuristic techniques can be used to optimize various business processes and functions within an enterprise. To be able to acquire valuable information from enterprise data, there is a need of metaheuristic methods usage, as this data is huge and complex to handle and analyse. The maximum set of raw enterprise data can be made by utilizing suitable ML-/DL-based model to perform metaheuristic optimization. The traditional optimization methods are inadequate to deliver optimum solution, as they terminate the search process, where the gradient value is close to zero. In contrast, metaheuristic algorithms can deliver solutions to mathematical characteristics for solving complex problems. Metaheuristics can be a valuable tool for optimizing various aspects of enterprise operations and decision-making processes, making them more efficient, effective and competitive in today's complex and dynamic business environment. The interpretation of three metaheuristic algorithms – PSO, GA and SMO – examined on the ten datasets. As the result Table 3.1 shows, PSO has the maximum average accuracy among the three algorithms, with a value of 91.85, while GA gives 91.49, and SMO shows 89.18. The predominant results show that PSO executes global and local search simultaneously, better in terms of iterations than GA. PSO has higher convergence rate as compared to GA and is able to avoid premature convergence to facilitate higher-quality solution. Therefore, PSO can be a popular choice for solving real-world problems.

In the future, the existing and hybrid metaheuristic algorithms can help to handle multi-objective optimization problems commonly encountered in enterprise development. They can be specifically designed to tackle large-scale optimization issues in enterprise system.

REFERENCES

1. Chrysanthos Maraveas, Panagiotis G. Asteris, Konstantinos G. Arvanitis, et al. "Application of Bio and Nature-Inspired Algorithms in Agricultural Engineering," *Archives of Computational Methods in Engineering*, vol. 30, no. 3, 2022.
2. Laith Abualigah, Amir H. Gandomi, Mohamed Abd Elaziz, et al. "Advances in Meta-Heuristic Optimization Algorithms in Big Data Text Clustering," *Electronics MDPI*, vol. 10, no. 2, 2021.
3. Kallol Biswas, Pandian M. Vasant, Jose Antonio Gamez Vintaned, et al. "A Review of Metaheuristic Algorithms for Optimizing 3D Well-Path Designs," *Archives of Computational Methods in Engineering*, vol. 28, pp. 1775–1793, 2020.
4. Mahdi Azizi, Siamak Talatahari, Amir H. Gandomi. "Fire Hawk Optimizer: A Novel Metaheuristic Algorithm," *Artificial Intelligence Review*, vol. 56, no. 1, 2022.
5. https://archive.ics.uci.edu/ml/datasets.php – UCI dataset.
6. Ahmed Elbeltagi, Nandlal Kushwaha, Jitendra Rajput, et al. "Modelling Daily Reference Evapotranspiration Based on Stacking Hybridization of ANN with Meta-Heuristic Algorithms Under Diverse Agro-Climatic Conditions," *Stochastic Environmental Research and Risk Assessment*, vol. 36, no. 10, pp. 3311–3334, 2022. Springer.
7. Ken Seier, Amol Ajgaonkar. "The Complete Guide to Computer Vision for Business Leaders," © 2021InsightDirect, USA.
8. Abdullah Ayub Khan, Asif Ali Laghari, Shafique Ahmed Awan. "Machine Learning in Computer Vision: A Review," *EAI Endorsed Transactions Scalable Information Systems*, vol. 8, no. 32, 2021.
9. Junyi Chai, Hao Zeng, Anming Li, Eric W. T. Ngai. "Deep Learning in Computer Vision: Acritical Review of Emerging Techniques and Application Scenarios," *Machine Learning with Applications*, vol. 6, p. 100134, 2021. ELSEVIER.

10. Muhammad Sufyan Khan, Farhana Jabeen, Sanaa Ghouzali. "Metaheuristic Algorithms in Optimizing DNN Model for Software Effort Estimation," *IEEE Access*, vol. 9, April 2021.
11. Prachi Agrawal, Hattan F. Abutarboush, Talari Ganesh. "Metaheuristic Algorithms on Feature Selection: A Survey of One Decade of Research (2009–2019)," *IEEE Access*, vol. 9, pp. 26766–26791, February 2021.
12. M. Castelli, L. Manzoni, L. Mariot, M. S. Nobile, A. Tangherloni. "Salp Swarm Optimization: A Critical Review," *Expert Systems with Applications*, vol. 189, p. 116029, 2021.
13. J. Del Ser, E. Osaba, J. J. Sanchez-Medina, I. Fister. "Bio Inspired Computational Intelligence and Transportation Systems: Along Road Ahead," *IEEE Transactions on Intelligent Transportation Systems*, vol. 21, no. 2, pp. 466–495, 2021.
14. J. Del Ser, E. Osaba, D. Molina, et al. "Bio-Inspired Computation: Where We Stand and What's Next," *Swarm and Evolutionary Computation*, vol. 48, pp. 220–250, 2019.
15. W. K. Wong, Chew Ing Ming. "A Review on Metaheuristic Algorithms: Recent Trends, Benchmarking and Applications," *Conference on Smart Computing & Communications*, 2019.
16. W. K. Wong, Chew Ing Ming. "A Review on Metaheuristic Algorithms: Recent Trends, Benchmarking and Applications," *International Conference on Smart Computing & Communications (ICSCC)*, 2019.
17. N. A. Kallioras, N. D. Lagaros, D. N. Avtzis. "Pity Beetle Algorithm–A New Metaheuristic Inspired by the Behavior of Bark Beetles," *Advances in Engineering Software*, vol. 121, pp. 147–166, 2017.
18. Muhammad Habib Rehman, Victor Chang, Aisha Batool, Teh Ying Wah. "Big Data Reduction Framework for Value Creation in Sustainable Enterprises," *Elsevier International Journal of Information Management*, vol. 36, no. 6, pp. 917–928, June 2016.
19. A. K. Kar. "Bio Inspired Computing–A Review of Algorithms and Scope of Applications," *Expert Systems with Applications*, vol. 59, pp. 20–32, 2016.
20. D. Dasgupta. *Artificial Immune Systems and their Applications*. Springer Science & Business Media, 2012.
21. X. Hu, Y. Shi, R. Eberhart. "Recent Advances in Particle Swarm," *Proceedings of the 2004 Congress on Evolutionary Computation (IEEE Cat. No. 04TH8753)*, vol. 1, IEEE, 2004.
22. Fred Glover. "Future Paths for Integer Programming and Links to Artificial Intelligence," *Center for Applied Artificial Intelligence, Graduate School of Business*, University of Colorado, Boulder, CO80300, U.S.A.
23. El-Ghazali Talbi, *Metaheuristics*. University of Lille–CNRS–INRIA, John Wiley & Sons, 2009.
24. S. Kirkpatrick, C. D. Gelatt, M. P. Vecchi. "Optimization by Simulated Annealing," *Science*, vol. 220, pp. 671–680, 1983.
25. Mohamed Abd Elaziz, Abdelghani Dahou, Laith Abualigah, et al. "Advanced Metaheuristic Optimization Techniques in Applications of Deep Neural Networks: A Review," *Springer, Neural Computing and Applications*, pp. 1–21, 2021.
26 L. M. Rasdi Rere, Mohamad Ivan Fanany. "Metaheuristic Algorithms for Convolution Neural Network," *Computational Intelligence and Neuroscience*, vol. 2016, 2016.

CHAPTER 4

Petrographic Image Classification Accuracy Improvement Using Improved Learning

Ashutosh Marathe, Tanuja Tewari and Falguni Vyas

4.1 INTRODUCTION

Image Processing finds applications in many multidisciplinary areas. Rock type detection is one such area. Pre-processing the acquired image followed by classification using pattern recognition principles is reported in [1–3]. The rock specimen mentioned therein were analysed macroscopically.

Assistive diagnostics using image processing and pattern classification over rock microstructures instead of handheld specimen is rarely done. The main reason being the challenging stochastic textures under consideration at microstructure level.

This chapter reports the research activity carried out to classify igneous rocks into volcanic and plutonic subcategories. Lava cooling rapidly on Earth's crust results in Volcanic rock formation. The fast-cooling results in small grain size of minerals. Plutonic rocks are formed below the earth surface. It is due to decelerated cooling of molten rocks, named magma. Owing to slow cooling, the minerals grow larger [4].

4.2 METHODS

The classification is carried out based on the mineral grain size. The analysis of textures is carried out based on approaches based on Haralick features [5] and Laws masks [6].

The stochastic nature of microstructures, poses a challenge that is not handled satisfactorily in terms of classification accuracy by single index or linear discriminant methods. The high complexity of the textures forces the disregard towards these methods. A non-linear SVM, such as radial basis function support vector machine (RBF SVM), random forests approach, naïve Bayes classifier approach, classifier combinatorial strategies typically provide adequate learning capacity and they demonstrate the ability to capture the

DOI: 10.1201/9781032699806-4

underlying complex non-linear models [7–9], which are a representative feature of a complex interdisciplinary study such as microstructure rock classification.

Although random forest classifier can achieve high accuracy in data classification compared with many standard methods, [10, 11] and it minimizes overall classification rate, the classification accuracy reduces with imbalanced dataset. Also, in the chosen IILA approach, run-time performance was important, hence random forest-based classification was not carried out.

Naïve Bayes classification is based on the principle of maximum posteriori hypothesis to identify the object that is most likely to be classified under that category [12]. However, the approach assumes independence between features, which is not the case with the given dataset.

Hence, the following classification algorithms were chosen as primary algorithms for iterative implementation:

1. RBF SVM classifier.
2. A further amendment is carried out based on fine tuning of controlling parameters C and Sigma are optimized using grid search [13]. Thus, an optimized RBF SVM was prepared and used for classification.
3. AdaBoost algorithm.
4. Convolutional neural networks

For the machine learning approaches, forward selection algorithm [14] is used for deciding the optimal dimensionality of the chosen feature set. The feature set is derived out of Haralick features and Laws masks.

The classification accuracy is calculated for each algorithm. Compared with a RBF SVM, AdaBoost algorithm registers highest percentage of accuracy for classification.

The 'improved learning' resulting due to addition of classified images into training set for the subsequent iteration improves the intra-algorithm percentage accuracy.

Further in this chapter, Section 4.3 gives the details of the database used. In Section 4.4, the details of feature pair selection are provided. Section 4.5 provides details of convolutional neural networks. Section 4.6 explains the basic concept of improved learning (IL). The description of the implementation of IILA approach is given in Section 4.7 with the help of a flowchart. Section 4.8 gives the details of results. Section 4.9 summarizes the conclusions derived out of the IL approach.

4.3 DATABASE

The research work has a local significance. The western part of India is based on igneous rocks. The conclusions derived about the presence of a volcanic or a plutonic rock based on the microstructure analysis has a direct local impact. Hence, the microstructure images representing igneous rock subfamilies belonging to volcanic and plutonic subclasses, such as andesite, pegmatites, basalt, spherulite, dolerite, pseudotachylites, rhyolites, etc., were selected. Various geological investigations carried out in northern India and western India confirm the abundance of these subfamilies [15–17]. The conclusions derived out of investigation of microstructures of these specimens are important for decision-making in the irrigation, civil work and construction activity.

The data base of 128 images - 64 each of plutonic and volcanic variety were selected. A proportionate presence of all subfamilies was ensured.

The challenge posed by the diverse stochastic textures were addressed by consideration of primary feature set on the basis of Haralick features and Laws masks.

The chosen Haralick features were homogeneity, correlation, energy, entropy and contrast. The Laws masks considered were absolute mean and standard deviation.

4.3.1 Methodology

The feature selection method adopted was forward selection, and the RBF SVM classifier was chosen.

4.4 FEATURE SELECTION

The objective of feature selection is to eliminate features that are not contributory to the classification. The objective is to improve generalization and expedite classification procedure compared with the process which can be realized with initial feature set.

Let the total set of selected features be F^m, where m is the number of input variables, and the recognition rate of the validation set by cross-validation be R^m.

Starting with the empty i.e. null set, the ith feature (i = 1, 2, . . . m) is augmented and the generalization ability is estimated by cross-validation.

The recognition rate of the validation set is R_i^m. It is checked whether the maximum R_i^m (i ε{1, ... m}) is larger than or equal to R^m.

$$\max R_i^m > R^m \text{ (for i = 1, . . . , m)} \tag{4.1}$$

If (1) is not satisfied, it is assumed that augmentation of the feature results in degrading the generalization ability; thus, we cannot add any further feature.

If (1) is satisfied for a certain k (k – 1, . . . , m),

Then set $F^i = F^m - \{k\}$.

This means that the set of features F^i can realize the same generalization ability as F^m.

To speed up selecting the features from F^i, considering that the features that satisfy $R_i^m < R^m$ are indispensable for classification and thus cannot be deleted.

Thus, the set of features that are candidates for selection are

$$S^i = \{i|\ R_i^m < R^m,\ i \neq k\}$$

If S^i is not full, features are continued to be augmented iteratively.

From a null set, progressive inclusion of features is carried out. Using a suitable classifier, the chosen classification metric is monitored. An optimum dimensionality feature set is identified. This is the forward selection method.

Per the binomial theorem,

$$\binom{n}{k} = \frac{n!}{k!(n-k)!} \tag{4.2}$$

Hence, the total number of ways of selecting distinct feature pairs out of 7 features per the binomial theorem was:

$$\frac{7!}{2!5!} = 21 \tag{4.3}$$

Classification performance was evaluated for the 21 duplets based on the average false rejection ratio (AFRR).

Composition of data for the trial: MATLAB® function randperm(n) was used to select half of the images of volcanic/plutonic rock. RBF SVM was applied, and the false rejection ratio (FRR) was calculated.

Based on 1000 trials, the AFRR was computed.

The AFRR for the pairs is as shown in Table 4.1.

The best AFRR value of 0.125 was registered for the absolute mean and contrast feature pair (bold emphasis).

TABLE 4.1 AFRR for 21 Duplets

Sr. No.	Duplet	AFRR
1	Homogeneity, correlation	0.32
2	Homogeneity, energy	0.26
3	Homogeneity, entropy	0.29
4	Homogeneity, contrast	0.355
5	Homogeneity, abs. mean	0.13
6	Homogeneity, std. deviation	0.18
7	Correlation, energy	0.34
8	Correlation, entropy	0.36
9	Correlation, contrast	0.275
10	Correlation, abs. mean	0.26
11	Correlation, std. deviation	0.31
12	Energy, entropy	0.26
13	Energy, contrast	0.21
14	Energy, abs. mean	0.31
15	Energy, std. deviation	0.26
16	Entropy, contrast	0.26
17	Entropy, abs. mean	0.31
18	Entropy, std. deviation	0.21
19	**Contrast, abs. mean**	**0.125**
20	Contrast, std. deviation	0.325
21	Abs. mean, std. deviation	0.24

Continuing with the forward selection method, triplets were formed with absolute mean and contrast, along with another feature. Out of the 5 triplets, triplet of absolute mean, contrast and energy provided the best AFRR of 0.175.

Table 4.2 summarizes the AFRR values obtained with addition of features.

The addition of features results in increase of computational time. However, this added dimensionality does not result in improvement of AFRR. Hence, the addition of features was deemed redundant, and the feature pair of absolute mean and contrast is considered.

4.5 CONVOLUTIONAL NEURAL NETWORKS

Convolutional neural networks comprise several neurons with learnable weights and biases stacked in a volume to form several layers. Mathematically, the output of each neuron in a convolutional layer is the response to some visual stimuli on the neurons of previous layer and hence is the result of convolutional operation on the neurons of the previous layer which lie in the receptive field of each next layer neuron. The visual stimuli are the learnable filter weights which exploit spatial locality. The architecture is designed such that the weights learnt produce the strongest response to spatially local correlation in images. Increasing the number of layers leads to increasingly non-linear filters which have indirect much larger spatial access. Hence, the architecture first creates low-level features which are ensembled in later layers for much larger representation [18].

4.5.1 Mathematical Interpretation of CNN

In statistics, the given data is given as (X_i, Y_i), where X_i is high-dimensional and $Y_i = f(X_i)$ is in {0,1} or $\mathbb{R}$. The goal is to find a function Y_i^* such that Y_i^* is close to Y_i for the predictions to be accurate.

In deep learning [19], these function Y_i are parametric and the family or class of functions is given by $f(X,\Psi)$ where Ψ (high dimensional) are the parameters for each layer. The parameters learnt Ψ^* are such that $f\left(X,\Psi^*\right)$is close to $f(X,\Psi)$. If the network comprises of d layers then the output function in the dth layer, K_j^d is a non-linear combination of n functions $\left(K_1^{d-1}, K_2^{d-1}, \ldots\ldots\ldots\ldots K_n^{d-1}\right)$ of $(d-1)^{th}$ layer.

TABLE 4.2 AFRR Values per Forward Selection Algorithm

No. of features	Features Involved	AFRR
2	Absolute mean and contrast	0.125
3	Absolute mean, contrast and energy	0.175
4	Absolute mean, contrast, energy and correlation	0.175
5	Absolute mean, contrast, energy, correlation and homogeneity	0.225
6	Absolute mean, contrast, energy, correlation, homogeneity and entropy	0.31
7	Absolute mean, contrast, energy, correlation, homogeneity, entropy and standard deviation (all seven features)	0.35

$f^{(d)}(.,\Psi)\circ\ldots\circ f^{(1)}(.,\Psi)$ are mostly high dimensional
$K_n(d\text{-}1)$

$K_1^{(d)}, K_2^{(d)}\ldots\ldots\ldots K_n^{(d)}$ are the components of the function $f^{(d)}$ in the layer d of the network. Also $K_j^{(d)}$ is the function of $\left(K_1^{d-1}, K_2^{d-1}, \ldots\ldots\ldots\ldots K_n^{d-1}\right)$. The function $f^{(d)}$ is taken to be non-linear. With linearity, there would be no use for multiple layers whose purpose is to introduce non-linearity to avoid underfitting. Hence, deeper the network, more would be the non-linearity introduced for complex classification. Hence, for the inputs $x_1, x_2, x_3 \ldots\ldots x_n \in \mathbb{R}_+$, the output of the model can be given as

$$f(x) = g\left(\sum_i \alpha_i x_i + c\right) \tag{4.4}$$

where $g(.)$ is the non-linear function.

Thus, deep neural networks comprise of the input layer and final output layer with several hidden layers in between. Every hidden unit in a hidden layer is the weighted combination of fixed number of hidden units in the layer below which is mapped to some deterministic value using a non-linear activation function.

$$K_j^{(l)} = g\left(b_j^{(l)} + \sum_i K_i^{(l-1)} w_{ij}^{(l)}\right) \tag{4.5}$$

where $K_j^{(l)}$ is the output of the jth neuron or hidden unit in l^{th} layer, $b_j^{(l)}$ is the bias associated with j^{th} neuron in l^{th} layer, and $w_{ij}^{(l)}$ denotes weights associated with j^{th} neuron in l^{th} layer operating on i^{th} neuron in $(l\text{-}1)^{th}$ layer.

This equation can be modified into vectorized form as $K^{(l)} = g^{\theta}(W^{(l)T}X + b^{(l)})$)for l^{th} layer whereas $g\theta$ denotes the coordinate-wise non-linear activation applied. $W^{(l)}$ and $b^{(l)}$ denote the learnable weights of the lth layer function component [19].

This notion can be extended to convolutional neural networks, which got its motivation from [20] that explains human visual cortex as a complex cell arrangement and those cells respond to only small visual field sub-sections. Hence, convolutional neural networks comprise of input layers, hidden layers and output layers. Input layer comprises of image samples to be classified, hidden layers comprise of the convolutional and pooling layers whereas output layer classifies the images to the respective classes that they belong. The input and the hidden layers are not fully connected. Each hidden layer neuron is the weighted combination of the small region of input neurons from the layer beneath it followed by an activation function. The output layer is fully connected. The performance of CNN depends on the amount of data fed to it for training. As evident from application areas, such as text [21], larger the amount of data better is the CNN's performance. The other drawback that CNN faces is the large amount of weights that need to be trained. Although the weights in a CNN are shared, i.e. the same set of weights are used for every sub-region in one layer, still with the increase in numbers of layers for complex classifications, the number of weights needed to be trained increase drastically, which makes the algorithm highly dependent on high-end expensive GPU machines for parallel computation [22].

4.5.2 Convolutional Layer

In convolutional layer, the weights are filters of much smaller size as compared to the image size. While calculating the activations for the next layer, they act on the smaller sub-region of the image or the current layer. Suppose, if there is a *N*×*N* square neuron layer followed by a convolutional layer. If *p*×*p* size filter comprising of learnable weights is used, the convolutional layer output will be of size $(N-p+1)\times(N-p+1)$. If, let's say, a nth *convolutional layer is represented by a grid, say m*×*m*, each position in it has local spatial connection with the following layer.

$K_{1,1}^{(n)}$	$K_{1,1}^{(n)}$	$\cdots$	$K_{1,m}^{(n)}$
$K_{1,1}^{(n)}$	$\ddots$		
$\vdots$		$\ddots$	
$K_{m,1}^{(n)}$			$K_{m,m}^{(n)}$

Hence, the output in the next layer $K_{ij}^{(n+1)}$ is the result of convolution of some fixed size grid say, *p by p* with the layer beneath it and a non-linear function *g*.

$$K_{ij}^{(n+1)} = g\left(\sum_{q=0}^{p-1}\sum_{r=0}^{p-1} w_{qr} K_{(i+q)(j+r)}^{(n)}\right) \tag{4.6}$$

where, $(n+1)$ is the current layer number; *i*, *j* are the row and column number respectively and w_{qr} are the weights of $(i, j)^{th}$ position associated with $(i+q, j+r)^{th}$ positions of nth layer.

If $p = 2$, the output of a neuron in $(n+1)^{th}$ layer is given as

$$K_{i,j}^{(n+1)} = g\left(w_{00}K_{(i,j)}^{(n)} + w_{10}K_{(i+1,j)}^{(n)} + w_{01}K_{(i,j+1)}^{(n)} + w_{11}K_{(i+1,j+1)}^{(n)}\right) \tag{4.7}$$

The weights for performing the convolution operation are updated on every iteration to achieve global minima of the loss function.

Hence, deep learning represents a class of hierarchical representation of signal followed by non-linearity at each hierarchy.

4.5.3 Pooling Layer

This layer has no weights to be learnt. The purpose of pooling layer is to downsample the feature maps. If *N*×*N* is the dimension of input layer, then output of pooling layer will be $\frac{N}{k}\times\frac{N}{k}$ layer. Every *k by k*block is downsampled by max pooling or average pooling to render a single value. Per the architecture required depending on resolution of image and other complexities, pooling layer follows one or many convolutional layers.

4.5.4 Output Layer

The final output layer comprises of several hidden units or neurons which are fully connected. The size of the output layer is equal to the number of classes to be classified.

4.5.5 CNN Training

Training the CNN implies, calculation of loss function at the end of feed-forward computation and backpropagating the loss to update the weights using an optimization algorithm. Weights are updated till an optimum value is reached where the cost function becomes minimum. Cost function is given as

$$J(W,b)=\left(\frac{1}{m}\right)\sum_{i=1}^{m}L(y^{*(i)},y^{(i)}) \tag{4.8}$$

where m are the number of training samples, $y^{*(i)}$ is the predicted posteriori probability, $y^{(i)}$ is the correct posteriori probability, and $L\left(y^{*(i)},y^{(i)}\right)$ is the loss function calculated over one training sample. The cost function $J(W, b)$ is a function of network weights W and bias b. Loss function is calculated by the cross entropy between the predicted and correct class posterior probabilities in classification problems.

$$L\left(y,y^{*};\Psi\right)=-\sum_{i}y_{i}\log\left(y_{i}^{*}\right) \tag{4.9}$$

Here Ψ stands for model parameters which comprise of weights and biases.

Backpropagation algorithm is used to propagate the errors back across the layers to compute the error gradients with respect to the weights. These error gradients are used to update the weights using optimization algorithms. Backpropagation is implemented in convolutional as well as the pooling layer.

4.5.6 Backpropagation in Convolutional LAYER

Using error function, E, the error values at the convolutional layer are computed. Gradient for each weight in the convolutional layer is computed by using errors of the convolutional layer. Calculation of the error values at the layer before the convolutional layer is calculated using the errors of the weights between the convolutional layer and itself. Mathematically the errors due to each weight can be calculated as follows:

$$\frac{\partial E}{\partial w_{ab}}=\sum_{i=0}^{N-m}\sum_{j=0}^{N-m}\frac{\partial E}{\partial x_{ij}^{l}}\frac{\partial x_{ij}^{l}}{\partial w_{ab}} \tag{4.10}$$

In this case, we must sum over all x_{ij}^{l} expressions in which w_{ab} occurs. This corresponds to weight-sharing in the neural network. By looking at the forward propagation equations,

$\frac{\partial x_{ij}^{l}}{\partial w_{ab}} = y_{(i+a)(j+b)}^{l-1}$. In order to compute the gradient, we need to know the values $\frac{\partial E}{\partial x_{ij}^{l}}$ (which are called 'deltas'). To compute the deltas using the chain rule:

$$\frac{\partial E}{\partial x_{ij}^{l}} = \frac{\partial E}{\partial y_{ij}^{l}} \frac{\partial y_{ij}^{l}}{\partial x_{ij}^{l}} \tag{4.11}$$

The error at the current layer is $\frac{\partial E}{\partial y_{ij}^{l}}$. To propagate errors back to the previous layer:

$$\frac{\partial E}{\partial y_{ij}^{l-1}} = \sum_{a=0}^{m-1} \sum_{b=0}^{m-1} \frac{\partial E}{\partial x_{(i-a)(j-b)}^{l}} \frac{\partial x_{(i-a)(j-b)}^{l}}{\partial y_{ij}^{l-1}} = \sum_{a=0}^{m-1} \sum_{b=0}^{m-1} \frac{\partial E}{\partial x_{(i-a)(j-b)}^{l}} w_{ab} \tag{4.12}$$

The dimension of these values is same as the layers to which they are backpropagated.

4.5.7 Backpropagation in Pooling Layer

The max-pooling layers do not actually do any learning themselves. Instead, they reduce the size of the feature maps by introducing sparseness. In forward propagation, *k*×k blocks are reduced to a single value. Then, this single value acquires an error computed from backwards propagation from the previous layer. This error is then just forwarded to the place where it came from. Since it only came from one place in the *k*×k block, the back-propagated errors from the pooling layers are less in number.

4.5.8 Algorithm

CNN algorithm is as follows.

Initial training and testing phase (first iteration):

1. Create training database by randomly selecting 32 samples from the plutonic rocks database and 32 samples from the volcanic rocks database using a random number generator. Thus, initially, the training database comprises 64 samples.
2. Apart from 32 samples each from the plutonic and volcanic databases for training, keep the rest 32 each from both the rocks databases, respectively, for testing. Thus, initially testing database too comprises of 64 samples.
3. Design the CNN architecture to be deployed for rock classification. Details of the model given in Table 4.3.
4. The training and test images are cropped to 64×64 *pixels* and normalized to feed as an input to CNN classifier.
5. Model is compiled using appropriate loss function and optimizer.
6. The training images along with the labels generated are fed to the classifier for training using appropriate *batch_size* and number of *epochs*.
7. Model is evaluated using the test database along with the known labels.
8. Keep a note of the training loss, training accuracy, testing loss and testing accuracy.

TABLE 4.3 Layers, Filter Size and Output Data Size Description of the CNN Model Input size of samples: 64×64

Layers Description	Filter Size Used	Padding Used, Stride Used	Size of Data
Conv layer	3×3×8	0,1	62×62×8
Relu activation			62×62×8
Conv Layer	3×3×8	0,1	60×60×8
Relu activation			60×60×8
Dropout			60×60×8
Conv layer	3×3×16	0,1	58×58×16
Relu activation			58×58×16
Conv layer	3×3×16	0,1	56×56×16
Relu activation			56×56×16 56×56×16
Dropout			
Conv layer	3×3×32	0,1	54×54×32
Relu activation			54×54×32
Conv layer	3×3×32	0,1	52×52×32
Relu activation			52×52×32 52×52×32
Dropout			
Flatten			86528
FC layer			256
Relu activation			256
Dropout			256
FC layer			2

9. Repeat *Steps* 1 *to Step* 8 for 100 iterations.
10. Calculate average test accuracy.

4.6 IMPROVED LEARNING

Balancing stability and plasticity is a concern which leads toa trade-off between maintaining acquired input while executing a task based on current data [23].

It is critical to strike the balance between plasticity and stability. Plasticity is required for a classifier requires plasticity to ensure that the new knowledge is integrated. At the same time, to prevent overlooking of the previous knowledge stability is equally important [24].

Compared to performance of many other classifiers based on stability, support vector machines are better performers [25].

Although stable, the lack of plasticity affects SVMs and are susceptible to catastrophic forgetting. It is an event, in which upon training a classifier for a certain task, followed by other result in overlooking the first task [26]. Therefore, to extract the full benefit from the SVM classifier, to retain its stability while ensuring plasticity, improved learning needs to be applied [27].

4.7 IMPLEMENTATION

For a given dataset, the overtraining of the classifier was to be avoided, or else it would perfectly learn the available data and fail on unseen data. Based on the alternatives for making best use of data, from amongst the re-substitution, hold-out, cross-validation and bootstrap methods, the hold-out method is selected since medium-sized data is being handled [28].

A random choice of half the images each from volcanic and plutonic datasets was made.

Ten thousand random combinations of training sets were generated and the one with the best percentage accuracy was further considered for iterative improved learning. The flowchart is given in Figure 4.1.

AdaBoost algorithm is as follows.

Training phase:

1. The parameters are initialized.

- Observe the weights $V^1=[V_1, \ldots, V_n], V_j^1 \in [0,1], \sum_{j=1}^{n} V_j^1 = 1$
- Consider the ensemble $P=\varphi$.
- Take M as the number of classifiers for training.

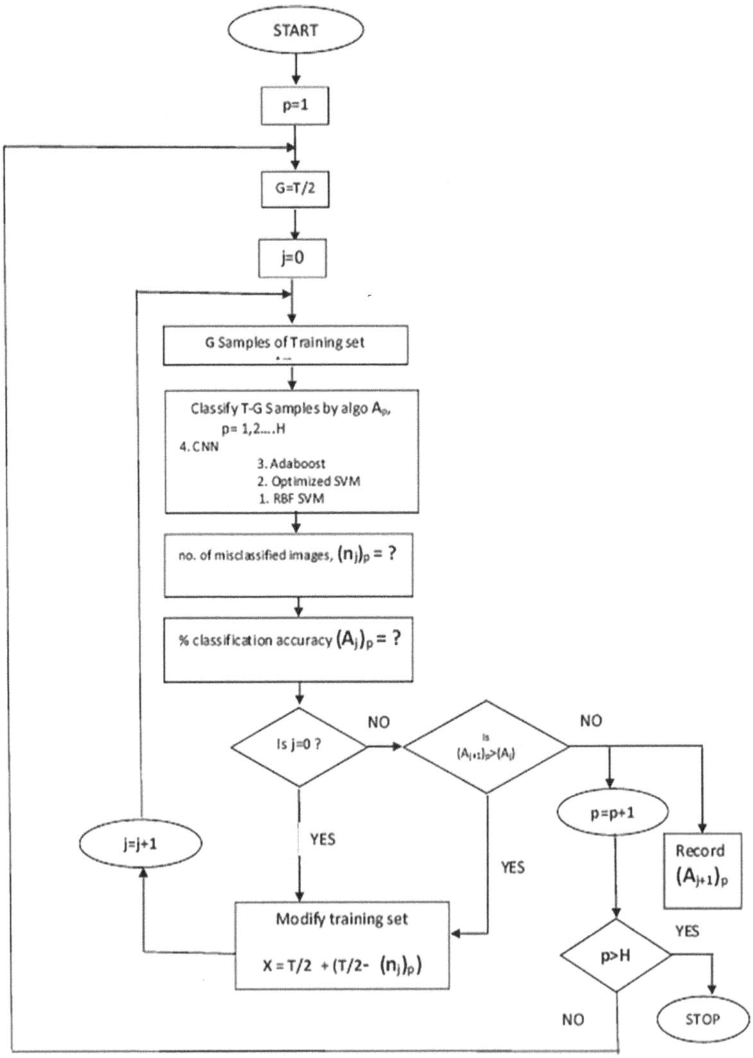

FIGURE 4.1 Implementation flowchart.

2. For q=1, . . . M

- Consider sample R_q from Z using distribution V^q.
- Develop a classifier P_q using R_q as the training set.
- Find the weighted ensemble error at step q by

  $\varepsilon q = \sum_{j=1}^{n} V_j^q l_q^j$,

  (l_q^j=1 if P_q misclassifies z_j and l_q^j=0 otherwise.)
- If ε_q=0 or $\varepsilon_q \leq 0.5$, then ignore P_k, the weights V_j^q are reinitialized to 1/n and the process is continued.
- Else find

  $\beta_q = \frac{\varepsilon_q}{1-\varepsilon_q}$ where $\varepsilon_q \in (0,0.5)$
- The individual weights are then updated

  $$V_j^{q+1} = \frac{V_i^q \beta_{al}(1-l_q^j)}{\sum_{i=1}^{n} V_i^q \beta_{al}(1-l_q^j)} \quad j=1,\dots n$$

3. Return P and β_1 and β_M $V_i^q \beta_{q\,(1-l}^{\ j}q)$.

Classification phase:

4. Find the support for class w_t by

$$\mu_t(y) = \sum_{Pq(y)=wt} \ln\left(\frac{1}{\beta q}\right)$$

5. Maximum support class is chosen as the label fory.

Improved learning algorithm:

1. Compute Classification Accuracy for Algorithm $A_v$with p = 1 for RBF SVM, p= 2 for Optimized RBF SVM [24] and p = 3 for AdaBoost having SVM as kernel classifier.
2. Record the misclassified images.
3. Mix the training images with the classified images from the iteration.
4. Iterate the revised training set
5. Classify the images misclassified in earlier iteration with the "revised" training dataset.
6. Record the details of wrongly classified images.
7. If reduction is seen in the wrongly classified images, repeat steps 1 to 6.

8. Else, stop.
9. The algorithm terminates when classification reaches its saturation and no further improved learning is possible.

4.8 RESULTS AND DISCUSSIONS

Table 4.4 provides the results for the given algorithm.

- **Result (a)**: For a given algorithm, as the learning improves, the percentage accuracy also improves. The AdaBoost classifier combination performs the best.
- **Result (b)**: The number of misclassified images progressively reduce, consistent with progressive improved learning.
- **Result (c)**: Some images that were misclassified with SVM algorithms, got classified with AdaBoost in first iteration, e.g. plutonic rock image no. 28. In further iterations, other images, such as plutonic image no. 47, also got correctly classified.

TABLE 4.4 Results

		(a) % Accuracy	
Algorithm	**Iteration 1**	**Iteration 2**	**Iteration 3**
RBF SVM	87.5	89.06	89.06
Opt. RBF SVM	89.06	92.18	92.18
AdaBoost	92.18	93.75	93.75
CNN	59.37	59.37	
		(b) No. of Misclassified images	
Algorithm	**Iteration 1**	**Iteration 2**	**Iteration 3**
RBF SVM	8	7	7
Opt. RBF SVM	7	5	5
AdaBoost	5	4	4
CNN	25	25	
		(c) Details of Misclassified images	
Algorithm	**Iteration 1**	**Iteration 2**	**Iteration 3**
RBF SVM	P8,29,30,34,40,47 V13,17	P8,29,30,40,47 V13,17	P8,29,30,40,47 V13,17
Opt. RBF SVM	P8,29,30,34,40,47 V13	P29,30,40,47 V13	P29,30,40,47 V13
AdaBoost	P29,30,40,47 V13	P29,30,40 V13	P29,30,40 V13
CNN	23 Plutonic and 2 Volcanic	23 Plutonic and 2 Volcanic	

4.9 CLASSIFICATION ERROR AND CLASSIFICATION ACCURACY

4.9.1 Apparent Error Calculation

For a labelled dataset Z_t having dimension N_t x n, where N_t is the dataset and n are dimensions of the feature set. Such dataset is used for testing the accuracy of the given classifier D_n.

The apparent error calculated for the given classifier D is given as

$$\text{Error (D)} = = \frac{1}{\mathbf{Nt}} \sum_{j=1}^{\text{Nt}} \left\{1 - \psi\left(l\left(Zj\right), Sj\right)\right\},\ \boldsymbol{Zj}\ \varepsilon Z_t,\ \text{where}$$

$\psi\left(\boldsymbol{a},\boldsymbol{b}\right)$)is an indicator function taking value 1, if a = b and is 0 if a ≠ b

4.9.2 Probabilistic Error calculation

The Classifier D commits an error with probability P_D on any object, then the number of errors has a binomial distribution with parameters (P_D and N_t)

An estimate of P_D is,

$P_{D\,=}\ N_{error\,/}\ N_t$ which is the computing error.

For the Igneous rock database,

N_t = 64, n = no. of features = 2

Sr. No.	Classifier No.	Classifier Algorithm	N_{error}	P_D
1	D_1	SVM	8	0.125
2	D_2	Optimized SVM	7	0.109375
3	D_3	AdaBoost	5	0.078125
4	D_4	CNN	25	0.390625

Since N_t and P_D satisfy the success Failure rule of thumb, N_t> 30, P_D x $N_t \geq 5$ and (1- P_D)x $N_t \geq 5$, the binomial distribution can be approximated as Normal distribution.

Suppose we take many training sets and build a confidence interval for each case, the 95% confidence interval for the error is given as

Point estimate ±1.96 x Standard Error (SE)

That is, in this case the 95% Confidence Interval Spread is

$$[Pd - 1.96\sqrt{\frac{Pd(1-Pd)}{Nt}}\ \text{to}\ Pd + 1.96\sqrt{\frac{Pd(1-Pd)}{Nt}}] \tag{4.13}$$

The 95% Confidence Intervals thus generated are as follows:

Sr. No.	Classifier No.	Classifier Algorithm	Confidence Interval
1	D_1	SVM	[0.044,0.206]
2	D_2	Optimized SVM	[0.033,0.185]
3	D_3	AdaBoost	[0.013, 0.143]
4	D_4	CNN	[0.271,0.510]

This means that whenever a relevant algorithm is implemented on randomly generated training sets, 95% of the time, the random sets will yield confidence intervals that contain the true population parameter, i.e. the classification error.

4.10 CONCLUSIONS

Percentage accuracy improves with improved learning. Amongst the algorithms used, AdaBoost algorithm performs the best. Volcanic rocks demonstrate smaller grain size compared to plutonic rocks. Based on the low contrast and low absolute mean, volcanic rock samples occupy bottom left position while the plutonic rock samples are on top left with slight overlap. The performance of AdaBoost was the best. The images which could not be classified with the other algorithms were classified with AdaBoost, e.g. plutonic image no. 8 and plutonic image no. 34.

IILA acts as a comprehensive procedure of database testing. It is used to indicate the typical hypabyssal "marginal" cases. The 95% confidence interval for the error is minimum for AdaBoost algorithm. The unique contributions of the research are (a) the development of improved learning algorithm; (b) its implementation for rock microstructure classification; (c) improvement in classification accuracy through combining classifier; (d) detection of the interim hypabyssal variety (which lies in the common portion shown in the volcanic and plutonic rock feature constellation) and its validation from geological perspective; and (e) the development of 95% confidence interval for error based on statistical rules for the three algorithms.

Convolutional neural networks, which are otherwise known to give state-of the art classification performance, give inferior results compared to other machine-learning-based classifiers when the training database is small since CNN's performance increases with the scale of data. On the other hand, in a case like this, where the training database is only 64 samples, to avoid underfitting, even if the system complexity is increased by using very deep CNNs with supporting expensive GPUs, it will suffer from the problem of high variance or system overfitting for the small training database and will not generalize well for unseen data. This create another dependency for performance of this classification algorithm on training data size. Thus, using CNN as a classifier for complex images as that of volcanic and plutonic rocks, although gives significant speed ups, yet seems to fail to achieve a trade off between system complexity, database size dependency and accuracy, which is better achieved by the proposed machine-learning-based algorithm with remarkable 93.75 % accuracy. The IILA implementation required 26.2 seconds on an Intel i3 processor working at 2.10 GHz, using MATLAB version R 2012 b. IILA had two iterations.

4.11 DECLARATIONS

1. **Availability of the data and material**: The data and material are available with the researchers, who are active members of the Centre of Excellence for Signal and Image Processing which is established under Technical Education Quality Improvement Program – Phase II (TEQIP-II), a World Bank initiative, at the place of Research

2. **Competing interests**: None.

3. **Funding**: The research work is carried out in Centre of Excellence for Signal and Image Processing, at the place of research. No earmarked separate funding under any research grant was utilized.

4. **Authors' contributions**: The authors have contributed in development of the algorithms, such as forward feature selection algorithm, algorithms for SVM, optimized SVM, AdaBoost SVM, and their implementation on the database. The authors have also contributed in providing the confidence-interval-based statistical basis for the accuracy calculation.

5. **Acknowledgement**: The authors would like to acknowledge the technical advisors of the Centre of Excellence for the guidance and suggestions from time to time.

REFERENCES

1. Shankar, V., Rodríguez, J.J., Gettings, M.E. 'Texture analysis for automated classification of geologic structures', *IEEE Southwest Symposium on Image Analysis and Interpretation*, 2006, pp. 81–85.
2. Zhou, Hang, Monteiro, S.T., Hatherly, P., Ramos, F., Nettleton, E., Oppolzer, F. 'Automated rock recognition with wavelet feature space projection and Gaussian Process classification', *IEEE International Conference on Robotics and Automation (ICRA)*, 2010, pp. 4444–4450.
3. Młynarczuk, Mariusz, Górszczyk, Andrzej, Ślipek, Bartłomiej. 'The application of pattern recognition in the automatic classification of microscopic rock images', *Computers & Geosciences*, vol. 60, pp. 126–133, July 2013. Elsevier Ltd.
4. Vernon, Ron. *A Practical Guide to Rock Microstructure*. New York, Cambridge University Press, 2004. https://researchers.mq.edu.au/en/publications/a-practical-guide-to-rock-microstructure.
5. Horlick, Robert, Shanmugam, K., Dinstein, Itzhak. 'Textural features for image classification', *IEEE Transactions on Systems Man Cybernetics*, vol. 3, no. 6, pp. 610–621, 1973.
6. Kenneth, Laws. 'Textured image segmentation', PhD Thesis. University of Southern California, Los Angeles, 1980.
7. Dong, L. et al. 'Discrimination of mine seismic events and blasts using the fisher classifier, Naïve Bayesian classifier and logistic regression', *Journal of Rock Mechanics Rock Engineering*, vol. 49, pp 183–211, 2016.
8. Dong, L. J., Li, X. B., Xie, G. N. 'Nonlinear methodologies for identifying seismic event and nuclear explosion using random forest, support vector machine, and naive Bayes classification', *Abstract and Applied Analysis*, vol. 2014, p. 459137, 2014.
9. Dong, L. J., Wesseloo, J., Yves, P., Li, X. B. 'Discriminant models of blasts and seismic events in mine seismology', *International Journal of Rock Mechanics and Mining Sciences*, vol. 86, pp. 282–291, 2016.
10. Sun, J., Zhong, G., Dong, J., Saeeda, H., Zhang, Q. 'Cooperative profit random forests with application in ocean front recognition', *IEEE Access*, vol. 5, pp. 1398–1408, 2017.
11. Lin, W., Wu, Z., Lin, L., Wen, A., Li, J. 'An ensemble random forest algorithm for insurance big data analysis', *IEEE Access*, vol. 5, pp. 16568–16575, 2017.
12. An, Y., Sun, S., Wang, S. 'Naive bayes classifiers for music emotion classification based on lyrics', *IEEE/ACIS 16th International Conference on Computer and Information Science (ICIS)*, 2017.
13. Yan, Z., Yang, Y., Ding, Y. 'An experimental study of the hyper-parameters distribution region and its optimization method for support vector machine with Gaussian Kernel', *IJSPIPPR*, vol. 6, no. 5, pp. 437–446, 2013.
14. Abe, Shigeo. *Support Vector Machines for Pattern Classification*. 2nd edition. London, UK, Springer, 2010.

15. Geological Survey of India Report. http://www.portal.gsi.gov.in/portal/page?_pageid=127,689645.
16. Agarwal, K. K. et al. 'Occurrence of pseudotachylites in the vicinity of South Almora thrust zone, Kumaun Lesser Himalaya', *Current Science*, vol. 101, no. 3, pp. 431–434, 2011.
17. Sisodia, M. S. 'Malani rhyolite: Highly eroded complex crater', *Current Science*, vol. 101, no. 7, pp. 946–951, 2012.
18. Lecun, Y., Bengio, Y., Hinton, G. 'Deep learning', *Nature*, vol. 521, no. 7553, pp. 436–444, 2015.
19. Srinivas Amiripalli, S., Nageshwara Rao, G., Behara, J., Sanjay Krishna, K. 'Mineral rock classification using convolutional neural network', In *Recent Trends in Intensive Computing*. Sanjivani College of Engineering, Kopargaon, India, IOP Press, pp. 499–505, 2021.
20. Hubel, D., Wiesel, T. 'Receptive fields and functional architecture of monkey striate cortex', *The Journal of Physiology*, vol. 195, no. 1, pp. 215–243, 1968.
21. Simard, P., Steinkraus, D., Platt, J. "Best practices for convolutional neural networks applied to visual document analysis', *7th IEEE International Conference on Document Analysis and Recognition (ICDAR)*, 2003, pp. 958–963.
22. Simonyan, K., Zisserman, A. 'Very deep convolutional networks for large-scale image recognition', *International Conference on Learning Representations*, 2015, pp. 1–14.
23. Modesire, S., Wiegand, R. 'A demonstration of stability-plasticity imbalance in multi-agent, decomposition based learning', *IEEE International Conference on Machine Learning and Applications (ICMLA)*, 2015, pp. 1070–1075.
24. Asaraf, S., Murty, M., Shevade, S. 'Rough set based incremental clustering of interval data', *Pattern Recognition Letters*, vol. 27, no. 6, pp. 515–519, April 2006.
25. Burges, C. 'A tutorial on support vector machines for pattern recognition', *Data Mining and Knowledge Discovery*, vol. 2, no. 2, pp. 121–167, 1998.
26. Goodfellow, I. J. et al. 'An empirical investigation of Catastrophic forgetting in gradient based neural network'. https://arxiv.org/pdf/1312.6211.pdf.
27. Kulkarni, Parag. *Reinforcement and Systemic Machine Learning for Decision Making*. Piscataway, NJ, Wiley – IEEE Press, 2012.
28. Kuncheva, L. *Combining Pattern Classifiers*. 2nd edition. Medford, MA, USA, John Wiley & Sons, 2014.

CHAPTER 5

Data Visualization and Dashboard Design for Enterprise Intelligence

Nishikant Bhaskar Surwade, Bahubali Shiragapur and Anwar Hussain

5.1 INTRODUCTION

5.1.1 Data Visualization: Key to Effective Enterprise Intelligence

Data visualization is graphically expressing information, data or insights. Enterprise intelligence utilizes data, technology and business processes to generate insights and support decision-making. Modern business intelligence requires data visualization since it makes it simpler to comprehend, analyse and communicate complex data clearly and concisely.

Companies are increasingly turning to visualizations to tackle their data management issues. Visual analysis improves data comprehension, which allows a broader range of users to understand the underlying patterns instead of relying solely on expensive data analysts. This expansion of access leads to new opportunities for decision-making, problem-solving, detecting trends and fostering creativity [1].

The effectiveness of data visualization in enterprise intelligence stems from its ability to deliver insights easily understood by all employees in an organization, regardless of technical skills. For example, a company might use a bar graph to show sales data, with the different bars representing different product lines. This simple way of showing the data is more accessible for everyone to understand than putting the same information in a spreadsheet or table, where it can be hard to see trends and patterns.

Another advantage of data visualization is that it can show vast amounts of information in a single view. Companies don't have to spend hours looking at spreadsheets and tables to make quick, well-informed decisions. Because data visualization is interactive, the user can look at the data and find insights in real time. This is helpful in healthcare, banking and retail areas, where many are created daily and quick decisions are needed.

Visualizing data is also essential for finding patterns and relationships in data. A scatter plot, for example, can show how the number of hours spent on a project and how much it

DOI: 10.1201/9781032699806-5

costs are related. This visualization makes it easier to see patterns and relationships, which can be used to make better decisions and grow a business. Additionally, data visualization can assist firms in identifying areas for improvement. A heat map, for example, can be used to depict customer satisfaction statistics, indicating which aspects of the business are working well and which require improvement. This visualization enables firms to make informed decisions regarding resource allocation and customer satisfaction enhancement.

In addition to its role in enterprise intelligence, data visualization is also vital for communication and collaboration. Clear and concise insight sharing with stakeholders inside and outside of the company is made possible by effective data visualization. This fosters credibility and trust, which can improve decision-making and teamwork. Data visualization can spur innovation and open up new doors. For example, you can use data visualization to find new markets, customer segments, and products and services. This can lead to new business models and growth prospects, which can be crucial to an organization's success in today's fast-paced and ever-changing business climate.

To summarize, data visualization is a powerful corporate intelligence tool that can help organizations make informed decisions, accelerate business success and communicate effectively with stakeholders.

5.1.2 Overview of Current Practices and Techniques

This section discusses data visualization and aesthetics. It defines data and information and describes classical data visualization, emphasizing data meaning for efficiency and effectiveness [2].

5.1.2.1 Charts

Using charts and graphs is one of the most common ways to show how data is organized. Bar charts, line charts, scatter plots and pie charts are just some charts and graphs that can be used to display data visually. Users can get dynamic interactive charts to help them interpret. These kinds of data visualizations are great for finding trends and patterns in data and sharing insights with stakeholders who may need more technical knowledge.

5.1.2.2 Heat Map

The usage of heat maps and geographical maps is another critical tool in data visualization. Heat maps are used to visualize data as a colour-coded map, with the intensity of the colour representing the data value. Geographical maps illustrate a description of the world or a specific region, using symbols, markers or shading to represent the data. These visualizations help find geographical trends and patterns in data, such as sales or customer satisfaction. This feature is also used widely in life science.

5.1.2.3 Interactive Dashboard

The usage of interactive dashboards and reports is another trend in data visualization. Dashboards and descriptions are intended to give consumers a rapid overview of key indicators while also allowing them to explore the data and discover insights in real time. This creates a business data management (DBM) that will enable users to interact with data by tracking, analysing, monitoring and displaying key business metrics. These dashboards can be tailored to an organization's specific requirements and visualize data from various sources, including spreadsheets, databases and cloud-based platforms.

5.1.2.4 Infographics

Another technique in data visualization is the use of infographics. Infographics are used to communicate complex data in a simple, easy-to-understand format. Infographics can be used to represent a wide range of data, from demographic data to sales data. They can be used to communicate insights to stakeholders who may need more technical expertise and can understand easily.

5.1.2.5 Visualization Using Augmented Reality

In addition to the methods that have been used for a long time, some newer methods are becoming increasingly popular. Creating virtual visualization is going to happen, which implants in mind events happening in the real world and enables human concentration, faith and belief. Augmented reality and virtual reality representations, possible to reduce design, production and maintenance costs, for example, are being employed to give consumers a more immersive experience. Augmented reality visualizations allow users to examine data in a virtual environment. In contrast, virtual reality visualizations provide a completely immersive experience that will enable users to view data in 3D, which provides the perception of depth, height and breadth of data information to help analyse data in qualitative and quantitative information of an object.

5.1.2.6 Using Machine Learning

Another data visualization trend is the application of machine learning and artificial intelligence. Machine learning techniques can be utilized to examine data and discover hidden patterns and correlations. Machine learning with statistical model approaches helps to understand the data. Artificial intelligence may generate intuitive insights and suggestions based on processed data. These techniques are beneficial in fields like finance, healthcare and retail, where much data is created daily and making quick decisions based on accurate information is critical to success.

To summarise, data visualization is an essential component of corporate intelligence, and substantial breakthroughs in data visualization processes and approaches have emerged in recent years. Data visualization, from basic charts and graphs to newer methods such as augmented reality and machine learning, offers companies the insights they need to make informed decisions and drive corporate success. Data visualization will become increasingly crucial in enterprise intelligence as data volume grows.

5.2 TECHNIQUES AND TOOLS

Data visualization is a critical aspect of enterprise intelligence, leveraging data, technology and business processes to provide insights and support decision-making. Several techniques and tools are available for data visualization in enterprise intelligence, and choosing the right ones depends on an organization's specific needs.

One common technique for data visualization is the use of charts and graphs. These can include bar charts, line charts, scatter plots and pie charts and help identify trends and patterns in data and communicate insights to non-technical stakeholders. Another popular technique is interactive dashboards and reports, which provide a quick overview of critical metrics and allow users to explore the data and discover insights in real time. These dashboards can be customized and integrated with various data sources, including spreadsheets, databases and cloud-based platforms.

Data visualization tools, such as Tableau, QlikView, Google Analytics and Power BI, are also widely used in enterprise intelligence, as they provide a platform for creating and sharing interactive visualizations and dashboards. These tools have a user-friendly interface, making it easy for non-technical users to create visualizations. They can also be integrated with other business intelligence tools and data sources. In addition to these traditional techniques, newer techniques, such as augmented reality and virtual reality visualizations, are becoming increasingly popular. These techniques provide a more immersive experience for users, allowing them to explore data in a virtual environment.

Machine learning and artificial intelligence are also used to analyse data and provide automated insights and recommendations. These techniques benefit organizations that generate large amounts of data and require quick, informed decision-making.

Several techniques and tools are available for data visualization in enterprise intelligence, and choosing the right ones depends on an organization's specific needs. From traditional charts and graphs to newer techniques such as augmented reality and machine learning, data visualization provides organizations with the insights they need to make informed decisions and drive business growth.

5.2.1 Taxonomy of Data Visualization Techniques and Tools

The taxonomy of data visualization techniques and tools refers to the classification and categorization of different methods and technologies used to display data. A taxonomy provides a clear understanding of the different types of visualization techniques available, as well as the features and benefits of each [3]. The following is a brief overview of the taxonomy of data visualization techniques and tools.

- **Statistical Visualizations**: Statistical visualizations are used to represent data to highlight the distribution, relationships and trends within the data. Examples include bar charts, line charts, histograms and scatter plots. These visualizations are particularly useful for analysing numerical data and identifying patterns or correlations.

- **Geographic Visualizations**: Geographic visualizations display data on a map and are often used to show spatial patterns, trends and relationships between variables. Examples include choropleth maps, dot density maps and heat maps.

  The primary goal of geographic visualizations is to display data on a map and help in uncovering spatial patterns and trends. For instance, choropleth maps are used to represent the distribution of a variable such as pollution levels or population density. They help in identifying the spatial patterns of the variable under study and can highlight areas with higher or lower values. Dot density maps could be used in displaying the spatial distribution of different species in an ecosystem. Finally, heat maps can be employed to represent the intensity and concentration of a particular phenomenon such as traffic congestion [4]. Therefore, geographic visualizations are an essential tool in spatial data analysis, facilitating the identification of patterns and relationships that may not be easily visible through other visualization methods.

- **Network Visualizations**: Network visualizations display data as nodes and edges and are often used to show relationships between objects, entities or people, such as social networks or supply chains. Examples include node-link diagrams, matrix visualizations and circular layout visualizations.
- **Temporal Visualizations**: Temporal visualizations display data over time and are often used to show trends and patterns in time-series data. Examples include line charts, bar charts and area charts. Temporal visualizations are critical tools for researchers examining time-series data [5]. These visualizations use graphical techniques to represent data over time and are effective in revealing trends and patterns in the data. Line charts are a common form of temporal visualization, where data points are connected by lines to show the evolution of the data over time. Bar charts are sometimes used to show changes in data over time, with individual bars representing each time period. Area charts are another popular method to display temporal data, where the area between the axis and the line is shaded to highlight the trend in the data. Hence, the effective use of temporal visualizations helps in communicating trends and patterns that occur over time in a clear and concise manner.
- **Hierarchical Visualizations**: Hierarchical visualizations display data organized in a tree-like structure and often show hierarchical relationships within the data. Examples include dendrograms, tree maps and sunbursts. Hierarchical visualizations are crucial to understanding complex data relationships in a variety of fields, including biology, marketing and finance. These visualizations display data in a tree-like structure, showcasing hierarchical relationships within the data. Dendrograms are a common form of hierarchical visualization utilized in the field of biology to explore phylogenetic relationships between organisms [6]. Tree maps display hierarchical data using nested rectangles, where each rectangle represents a category and is proportionate to its value. Sunbursts, on the other hand, are used to visualize multi-level and hierarchical data, where each ring represents a different level of organization. Hence, hierarchical visualizations provide an excellent means to represent hierarchical relationships within data for easy interpretation.
- **Interactive Visualizations:** Interactive visualizations allow users to manipulate and explore the data and provide an interactive experience for discovering insights. Examples include dashboards, drill-downs and detail-on-demand visualizations.

  Interactive visualizations allow users to engage with data and explore the information in a dynamic and intuitive manner. These visualizations provide an interactive experience for discovering insights, as they allow users to manipulate and explore the data. Dashboards are a popular example of interactive visualizations that provide high-level summaries of complex datasets, allowing users to view and analyse key metrics in real time. Drill-downs allow users to explore visualizations at a more granular level, providing detailed views of the data [7]. Hence, interactive visualizations offer a user-driven approach to discovering insights that makes data analysis more approachable to a broader audience.

- **Immersive Visualizations:** Immersive data visualization has become a topic of great interest in recent years, as it provides a more engaging and interactive experience for users compared to traditional 2D graphs and charts [8]. Immersive visualizations use virtual and augmented reality to create engaging and interactive environments for exploring complex data. For example, virtual reality visualizations enable users to immerse themselves in a virtual environment to explore and analyse data, leading to a better understanding of complex datasets [9]. Augmented reality visualizations, on the other hand, use real-world environments with data overlays to provide contextual information to users [10]. Immersive data visualization has great potential for enterprise intelligence as it offers a more intuitive and engaging approach to data analysis.

In terms of tools, there are several popular tools for data visualization, including Tableau, Power BI, QlikView, D3.js, fusionCharts and Matplotlib, to name a few. Each device has its features and benefits, and choosing the right tool depends on an organization's specific needs.

5.2.2 Comparison of Different Techniques and Tools

Comparing different data visualization techniques and tools is essential in choosing the right approach for an organization. While each method and tool has its strengths and weaknesses, the following are some key considerations when comparing different options.

- **Type of Data:** The type of data being visualized is a crucial consideration, as some techniques and tools are better suited for specific types of data, such as time-series data, geographic data or hierarchical data. For example, time-series data requires charts that can effectively show trends over time, such as line charts or area charts. On the other hand, geographic data may require maps or heatmaps to effectively show geographical patterns and anomalies [11]. Similarly, hierarchical data may require tree maps or sunburst charts to effectively show the hierarchical structure of the data [12]. Therefore, the type of data being visualized plays a crucial role in the selection of the appropriate chart type or visualization tool.
- **Use Case:** When considering the use case for visualization, it's crucial to choose appropriate techniques and tools, as each serves a unique purpose. For instance, dashboard design requires a simplified visual representation of complex data to help convey critical information for making business decisions [13]. Data exploration, on the other hand, uses interactive and dynamic visualizations to reveal underlying patterns and trends in the data [14]. Finally, reporting involves communicating insights

derived from data through a visually rich story [15]. Therefore, selecting the suitable technique and tool based on the use case can determine how clearly and effectively the information is communicated to the stakeholders.

- **User Skill Level**: The skill level of the users who will create and use the visualization is also a key consideration, as some techniques and tools may be more user-friendly than others. For example, tools such as Tableau or Power BI offer a user-friendly, drag-and-drop interface that allows novice users to create impactful visualizations easily [13]. In contrast, tools such as R or Python require coding skills and may be better suited for users with more advanced technical skills [16]. Therefore, the skill level of users is an important consideration in selecting the appropriate visualization technique and tool, as it can impact how efficiently and effectively the visualization is created and utilized.
- **Interactivity:** The level of interactivity desired in the visualization is also necessary, as some techniques and tools may provide more interactive features than others. Interactive features can include zooming in and out of data, filtering and drilling down to specific details. Tools such as D3.js or Vega-Lite provide advanced interactive features for complex visualizations [17]. In contrast, tools like Tableau provide a user-friendly and interactive interface but offer limited customization options for advanced interactive features [13]. Therefore, considering the level of interactivity the user desires is vital to achieve maximum user engagement and effective communication of insights.
- **Integration:** In today's data-driven environment, the ability to integrate visualizations with other tools and data sources is crucial. Integration enables users to combine insights from multiple sources to obtain a more comprehensive picture of the problem. Tools such as Power BI and Tableau offer excellent integration capabilities, with the ability to connect to different data sources and APIs. Therefore, selection of a visualization technique and tool that offers a high level of integration with other data sources and platforms is vital to gaining maximum insights from available data.
- **Cost:** The cost of the technique or tool is also an important consideration, as some options may be more expensive than others. The cost of a data visualization technique or tool is a significant factor to consider, particularly for small to medium-sized enterprises. While proprietary tools provide advanced features and functionalities, they may require a considerable initial investment. Open-source visualization tools may offer a lower cost alternative with sufficient capabilities and customization options, making them a viable choice for cost-conscious organizations. Therefore, selecting a visualization technique or tool that aligns with the budget and requirements of the enterprise is crucial for effective decision-making.
- **Scalability:** Scalability is an essential consideration when selecting a data visualization technique or tool for enterprise intelligence. As datasets grow larger and more complex, some visualization tools may struggle with rendering and processing the data. For instance, traditional charting tools may become slow and unresponsive

when handling large amounts of data. In contrast, specialized tools may offer better scalability features with efficient algorithms and optimized architecture, making them better suited for enterprise-scale data visualization requirements. Therefore, it is important to assess the scalability of the selected tool to ensure it can handle increasing amounts of data in the future.

When comparing different techniques, it is essential to consider the type of data being visualized, the use case for the visualization, the skill level of the users, the level of interactivity desired, the ability to integrate with other tools and data sources, the cost and the scalability.

In terms of tools, Tableau and Power BI are often compared as they are both popular and widely used for data visualization in enterprise intelligence. Tableau is known for its versatility and ability to handle large amounts of data, while Power BI is known for its ease of use and integration with the Microsoft ecosystem. QlikView is also a popular option, known for its fast performance and ability to handle complex data. The right visualization technique and tool can help organizations unlock insights and drive business growth.

5.2.3 Current Methods and Techniques

Data visualization has come a long way in recent years, and many different methods and techniques are available for displaying data in a meaningful and impactful way.

The following is a brief overview of some current data visualization methods and techniques.

- **Data Visualization Dashboards:** Dashboards are a popular method for displaying data and are often used to provide an at-a-glance view of critical metrics and KPIs (key performance indicators). Dashboards can be customized to display data in a way that is meaningful to the user and can be designed to be interactive and allow for drill-downs into more detailed data. According to the IDC report, dashboards and scorecards remain one of the most popular methods for displaying data in the business intelligence (BI) market [18]. Dashboards can be tailored to visualize data in ways that make sense to the user, and it offers an opportunity to interact or drill-down into more granular data. Additionally, dashboards provide a means to track progress towards a specific target with each indicator assigned to a vital objective, turning the data into actionable insights for the business [19].
- **Interactive Visualizations:** Interactive visualizations provide users with the ability to manipulate and explore the data, resulting in more user engagement and effective data communication. Examples of such visualizations include dashboards, drill-downs and detail-on-demand visualizations. A study by Wu and Chen [20] highlights the significance of interactive visualizations in enhancing user engagement and understanding as compared to static visualizations. Moreover, combining interactive visualizations with other data visualization tools can result in improved analysis and decision-making [21]. Therefore, it's crucial to incorporate interactive visualizations in data visualization and dashboard design to provide a more interactive experience for users.

- **Geospatial Visualizations:** Geospatial visualizations display data on a map and are often used to show spatial patterns, trends and relationships between variables. Examples include choropleth maps, dot density maps and heat maps. Geospatial visualizations are often utilized to examine spatial data patterns for commercial purposes, such as tracking the number of customers per location or product sales in various states or countries. They also offer vast benefits for exploratory data analysis (EDA), such as finding cluster formations and understanding relationships between variables in different areas [22]. Therefore, geospatial visualizations can provide valuable insights into spatial data and help businesses make informed decisions.
- **Network Visualizations:** Network visualizations are used to display data as nodes and edges and are often used to show relationships between objects, entities or people. Examples include node-link diagrams, matrix visualizations and circular layout visualizations. Network visualizations can be of great help in data exploration and analysis, network representation and social media analysis [23]. These visualizations provide an overview of the connections between entities and highlight the strength and weakness of relationships. As such, network visualizations serve as an excellent decision-making tool by depicting how entities are interlinked and organized across an organization or a network.
- **Temporal Visualizations:** Temporal visualizations display data over time and are often used to show trends and patterns in time-series data. Examples include line charts, bar charts and area charts. Temporal visualizations allow users to track change over time and identify trends or patterns, making them useful in various domains, such as finance, biology and economics [24]. For instance, a line chart can be used to exhibit the fluctuation of stock prices over a particular time period. In summary, temporal visualizations provide a valuable way to display complex data and help users to navigate a detailed analysis of patterns over time.
- **Immersive Visualizations:** Immersive visualizations use virtual and augmented reality to provide a more immersive experience for exploring data. Examples include virtual reality visualizations and augmented reality visualizations. These visualizations immerse users to interact with an environment that simulates the real world, with the ability to examine volumetric data, explore large datasets and visualize data in 3D [25]. Immersive visualization has applications in various domains, such as telemedicine, architecture and education. For instance, virtual reality visualizations can be used to visualize 3D anatomical structures in the healthcare domain. Overall, immersive visualization techniques offer a new level of immersive experience, enabling users to comprehend complex data with high interactivity.
- **Statistical Visualizations**: Statistical visualizations are used to represent data to highlight the distribution, relationships and trends within the data. Examples include bar charts, line charts, histograms and scatter plots. Statistical visualizations allow users to gain insights into complex datasets and communicate their findings more effectively. For example, a scatter plot can be used to visualize the relationship

between two variables, such as the correlation between height and weight. In summary, statistical visualizations offer a powerful way to identify patterns, relationships and communicate data insights.

In recent years, technological advancements have allowed for more advanced and sophisticated data visualizations, such as immersive and interactive visualizations. As data plays an increasingly important role in decision-making, the methods and techniques for visualizing data will continue to evolve and improve. There are many different methods and techniques for data visualization, and the right approach depends on an organization's specific needs. From dashboards to immersive visualizations, the correct method and technique can help organizations unlock insights and drive business growth.

5.3 DATA VISUALIZATION IN ACTION

5.3.1 Industry Application

Data visualization is used in various industries to help organizations make better decisions and gain valuable insights from their data. Here are a few examples of how data visualization is being used in different industries:

- **Healthcare**: Data visualization plays a vital role in the healthcare industry, enabling healthcare providers to make informed decisions to improve the quality of care provided to patients. Data visualization is an essential tool used in healthcare for analyzing and communicating complex data. It enables healthcare providers to derive insights that can improve patient outcomes and public health initiatives [26]. For instance, public health officials use visualization tools to track disease outbreaks and monitor trends in the spread of diseases [27]. Data visualization also helps doctors to understand patient outcomes and identify areas of improvement in service delivery (Peleg, Shahar, & Quaglini, 2017). Furthermore, visualization tools can help healthcare providers to visualize and analyze large volumes of medical data, such as electronic health records (EHRs), to improve data management and accuracy [28].
- **Finance**: In finance, data visualization is used to track financial performance, monitor market trends and make informed investment decisions. Visualization tools can help financial institutions to analyze large amounts of data and make informed decisions based on real-time information. Visualization also helps in identifying trends and patterns in the market, providing a better understanding of the overall financial climate. For instance, studies show that visualizing stock market trends enable investors to make informed decisions and increase the chances of getting high returns on investments [29]. Furthermore, visualization tools for portfolio analysis have been found to be effective in risk management, helping to identify high-risk areas and reducing the probability of loss [30].
- **Retail**: Data visualization tracks sales trends, monitor customer behaviour and optimizes supply chain management. Visualization tools can help retailers to identify patterns and trends in their data, allowing them to make informed decisions about inventory management, pricing and marketing strategies. This improved visibility and

understanding of user preferences aids retailers in gaining better insights into supply chain management activities. These insights help retailers enhance their inventory management process, automatically calculate key metrics that fluctuate rapidly, such as inventory turnover and reveal optimal times to place orders for products. By optimizing supply chain management, retailers can lower transportation and warehousing costs, reduce out-of-stock events and improve customer satisfaction levels [31].

- **Manufacturing**: In manufacturing, data visualization is used to track production performance, monitor machine efficiency and optimize supply chain management. Visualization tools can help manufacturers to identify patterns and trends in their data, allowing them to make informed decisions about production processes, resource allocation and inventory management. For instance, the integration of real-time monitoring systems and IoT devices with visualization dashboards has enabled manufacturers to detect production irregularities and take corrective measures in real time, leading to cost reductions and improved operational efficiency. Likewise, visualization tools facilitate effective inventory management, enabling manufacturers to analyse procurement data with a long-term vision and reduce stock-outs significantly. By harnessing data visualization technology, manufacturers can improve production planning, eliminate waste in resource allocation and ultimately create better products with higher-quality outputs.
- **Energy**: In the energy industry, data visualization is used to track energy usage, monitor renewable energy sources and optimize energy production. Visualization tools can help energy companies to identify patterns and trends in their data, allowing them to make informed decisions about energy production, resource allocation and sustainability initiatives.

  The industry is data-intensive and generates a vast amount of data that needs to be analysed, managed and shared for multiple purposes. With data visualization tools, energy companies can explore and analyse complex datasets to identify patterns and insights that are difficult or impossible to discern with traditional analytics approaches. For instance, using time-series visualizations, analysts can track energy usage across different regions and timeframes, identify peak demand periods and optimize energy production accordingly. They can also track and monitor renewable energy sources, such as solar, wind and geothermal power, and assess their contribution to energy output. Besides, data visualization can help energy companies to monitor carbon emissions and develop sustainability initiatives
- **Education**: To track the progressive performance of students based on continuous evaluation for various rubrics. The interactive dashboards will help mentors to know a child's progress at an early age and may help identify students' intelligence and concentration levels. Continuous evaluation enables the measurement of a student's progress over time based on various rubrics. Interactive dashboards can be used by mentors to track children's progress and spot any performance deficiencies early on, enabling them to devise custom learning plans. This approach can also help identify students' intelligence and concentration levels, resulting in more efficient classrooms and improved student outcomes.

- **Transportation**: Data visualization is an important tool in the transportation industry, used for monitoring and optimizing vehicle performance and routes. Visualization of GPS data enables route optimization for efficient delivery schedules. Additionally, real-time monitoring of traffic flow using visualization tools allows transportation companies to reduce traffic congestion and improve safety [32]. Data visualization also enables transportation companies to identify patterns and trends in their data, helping them make informed decisions about resource allocation, route optimization and delivery schedules [33]. For instance, visualization of vehicle speed data can reveal areas of congestion and allow for better route planning. Therefore, data visualization plays an important role in the transportation industry, enabling companies to increase efficiency, improve safety and reduce costs.

Data visualization is used in various industries to help organizations make better decisions and gain valuable insights from their data. From healthcare to transportation, the right data visualization tools can support business growth and drive success

5.3.2 Data Visualization: Industry Results and Effectiveness

The effectiveness of data visualization in various industries is highly dependent on the specific use case and the tools used. However, data visualization has been shown to impact many industries positively and has become an essential tool for making informed decisions.

- **Healthcare**: In healthcare, data visualization and blockchain tools be effective in tracking patient outcomes, monitoring disease outbreaks and analysing trends in health data. By visualizing complex datasets, healthcare providers can identify patterns and relationships that can be used to improve patient outcomes and support public health initiatives. The data visualization tools have been used in identifying and managing chronic diseases by examining large amounts of health records, which enable healthcare practitioners to identify patients who require more intense monitoring and to track patient outcomes. Also, data visualization enhances patient engagement by empowering patients with a better understanding of their health status through accessing their health record data.
- **Finance**: Data visualization with fintech tools effectively tracks financial performance, monitors market trends and makes informed investment decisions. By visualizing financial data in real time, financial institutions can make informed decisions based on up-to-date information and gain valuable insights into market trends and performance.

  Real-time visualization of financial data has proved to be crucial for financial institutions as it allows for up-to-date information on market trends and performance. For example, visualization of stock market data has been used to monitor market trends and identify profitable investment opportunities [34]. Moreover, data visualization has facilitated the tracking of financial performance by enabling the identification of patterns and visualizing financial reports [35]. The utilization of visualization tools in the finance industry is therefore essential for informed decision-making and for gaining valuable insights into market trends and performance.

- **Retail**: Data visualization has become increasingly important in the business world, particularly in industries such as retail where its effectiveness has been proven. One study by Business Intelligence Reports found that companies using data visualization tools were more likely to find timely information than those that relied on traditional reporting methods [13]. In the retail industry specifically, data visualization has enabled businesses to track sales trends, monitor customer behaviour and optimize supply chain management. By visualizing data on sales trends, customer behavior and inventory management, retailers can make informed decisions about pricing, marketing strategies and resource allocation. Moreover, the use of AI-based neuro-marketing has allowed for more accurate monitoring of customer behavior, further enhancing retailers' ability to drive business outcomes [36]. These examples highlight how data visualization can enhance decision-making, improve communication and ultimately drive business outcomes in the retail industry.
- **Manufacturing**: In manufacturing, data visualization is effective in tracking production performance, monitoring machine efficiency and optimizing supply chain management. By visualizing production data, manufacturers can identify patterns and trends in their data, allowing them to make informed decisions about production processes, resource allocation and inventory management.
- **Energy**: In the energy industry, data visualization is effective in tracking energy usage, monitoring renewable energy sources and optimizing energy production. By visualizing energy data, energy companies can identify patterns and trends in their data, allowing them to make informed decisions about energy production, resource allocation and sustainability initiatives.
- **Transportation**: In transportation, data visualization is effective in tracking vehicle performance, monitoring transportation routes and optimizing delivery schedules. By visualizing transportation data, transportation companies can identify patterns and trends in their data, allowing them to make informed decisions about resource allocation, route optimization and delivery schedules.

Data visualization results and effectiveness in various industries have been largely positive. By visualizing complex datasets, organizations can identify patterns and relationships in their data that can be used to make informed decisions and drive business growth. However, data visualization's specific results and effectiveness can vary depending on the use case and tools used [37].

5.4 FRAMEWORKS FOR DATA VISUALIZATION

Several frameworks are available for implementing data visualization in enterprise intelligence, each with its strengths and weaknesses. Some of the most commonly used frameworks include the following:

- **D3.js:** The field of data visualization has seen a significant shift towards building data visualization frameworks that cater to enterprise intelligence needs. One such framework is D3.js, which allows developers to create highly engaging, browser-based visualizations with complete control over the styles, layouts and interactivity of data visualization. D3.js is a JavaScript library for producing dynamic, interactive data visualizations in web browsers. It uses HTML, CSS and SVG to create powerful visualizations that can be easily integrated into web applications. One of the strengths of D3.js is its ability to handle large amounts of data and provide highly customizable visualizations [38].

  In recent years, D3.js has seen widespread adoption and has become the de facto standard for creating interactive and scalable visualizations on the web. The framework's power lies in its ability to handle large volumes of data and customize visualizations based on specific requirements. Kulkarni et al. [39] state that D3.js visualizations are high-performance, responsive and dynamic, enabling users to perceive and comprehend complex data patterns through immersive experiences. D3.js can simplify complex models and equations to a more comprehensive visual representation, making it easier for end-users to understand key trends and patterns and make informed decisions. With its consistent evolution and a substantial community of developers, D3.js is well-positioned to be a major contributor to enhancing enterprise intelligence through data visualization.

- **Tableau**: Tableau is a data visualization software organizations use to explore, analyse and communicate their data. It provides a user-friendly interface for creating interactive visualizations, dashboards and reports. Tableau is widely used in enterprise intelligence due to its ability to handle large amounts of data, support a wide range of data sources and integrate with popular data warehousing tools [40]. In addition, the software can handle large amounts of data, support a vast range of data sources and even integrate with popular data warehousing tools. Tableau's user-friendly interface makes it a popular choice for businesses without trained data analysts, allowing them to create data visualizations and dashboards that can guide their decision-making processes. Additionally, Tableau's interactive visualizations help to identify data trends, discrepancies or anomalies, and predictions which aid in the proper management of businesses.

- **Power BI**: Power BI is a business intelligence and data visualization tool developed by Microsoft. It provides a user-friendly interface for creating interactive visualizations, dashboards and reports. Power BI is widely used in enterprise intelligence due to its ability to handle large amounts of data, support a wide range of data sources and integrate with popular data warehousing tools [41]. Power BI is one of the leading data visualization frameworks that are widely adopted in enterprise intelligence. Its popularity is attributed to its ability to provide businesses with powerful visual data analysis tools that support real-time monitoring, KPI tracking and predictive analytics. In addition, Power BI's integration with various data sources, such as Excel, SharePoint and Dynamics 365, makes it an ideal choice for organizations that have disparate data sources. Furthermore, recent research has shown that Power BI's natural language query engine provides a faster and more intuitive way to obtain insights from data [42].

- **QlikView**: One of the frameworks that have gained popularity is QlikView. QlikView incorporates associative search technology that allows users to navigate through data in a way that traditional query-based approaches cannot provide. It provides an easy-to-use interface for creating interactive dashboards and visualizations. QlikView can handle large volumes of data and integrates seamlessly with various data warehouses, including SAP, Oracle and Microsoft SQL Server. Moreover, QlikView uses an in-memory data model that speeds up the query response time [43, 44]. The flexible data model permits real-time analysis, linear scalability and self-service business intelligence. The platform comes with robust security features that ensure secure access to data within an organization. QlikView has found applications in various fields, including healthcare, finance and manufacturing, and through its comprehensive range of features, QlikView has become a top data visualization tool for enterprise intelligence. The integration of QlikView into an organization results in faster and more accurate decision-making.
- **ggplot2**: One popular data visualization library used in enterprise intelligence is ggplot2. ggplot2 is a part of the R programming language and has gained popularity due to its flexibility in creating highly customizable visualizations for handling information-dense data. ggplot2 uses a layered grammar of graphics approach that allows users to combine different visual elements to create complex and informative visualizations. It is widely used in diverse fields such as finance, healthcare and marketing due to its ability to integrate with popular data warehousing tools. By efficiently handling and parsing large datasets in real time, ggplot2 enables organizations to derive valuable insights from their data, resulting in better decision-making [45, 46].

Several frameworks are available for implementing data visualization in enterprise intelligence, each with its strengths and weaknesses. The best framework for a particular organization will depend on the requirements, data sources and tools used. The most commonly used frameworks include D3.js, Tableau, Power BI, QlikView and ggplot2.

5.4.1 Research on Data Visualization Frameworks

The research by Noonpakdee et al. presents a framework for analysing and creating dashboard templates for small and medium-sized enterprises. The focus is on sales templates deemed simple for SMEs to understand. The framework includes four elements: business operation, attribute list, visualization and dashboard features. Forty SMEs used the templates for six to nine months, and the study evaluated their performance, benefits, challenges and potential improvements [47].

Zimmerman et al. highlight that the need for sufficient data visualization in current social media tools limits the ability to extract meaningful insights and make informed decisions from social data [48]. This chapter presents the technical design of a social business intelligence (BI) prototype tool, Social Newsroom, under development. The authors use an action design research approach to provide practitioners with user-friendly interfaces for leveraging social data. This chapter outlines the construction of specific interfaces, such as monitoring dashboards and insight pillars for visual analytics [48].

Harrison et al. studied the appeal of infographics and found that people quickly form a reliable first impression based on colourfulness and visual complexity. Age, gender and education also impacted the preferred level of colourfulness and complexity. These findings suggest that the outcomes of engagement and memorability are determined quickly [49].

The framework proposed by Mendoza et al. presents a four-stage process for creating effective and compelling dashboards. The stages include data preparation, prototype, revision and publishing and the framework also includes tips for good dashboard design, such as focusing on a clear message and avoiding redundant information, as well as suggestions for colour use and layout. The authors highlight that by following the framework, developers can create dashboards that are visually appealing, easy to use, and effective in conveying the intended message [50].

Moreover, the research by Hullman et al. argues that simple visualizations, such as bar charts, can be more effective than more complex ones for communicating data, especially when the goal is to compare quantities. The authors suggest that designers should consider task requirements and user familiarity with specific types of visualizations when selecting an appropriate design. They also observe that creating multiple, complementary visualizations can provide a more comprehensive view of the data, resulting in a more nuanced understanding of the information [51].

5.4.2 Key Considerations for Selecting Right Framework

Choosing the appropriate technique and tool for a specific scenario is a critical step in data visualization. The following factors should be considered when making this decision:

- **Data Type**: The first step in choosing the appropriate technique and tool is to consider the data type that will be visualized. Different techniques and tools are better suited for data types, such as time-series, categorical or geographical data.
- **Data Volume**: Visualizing the amount of data is also essential in choosing the appropriate technique and tool. Some tools may need help to handle large amounts of data, while others may be optimized for large datasets.
- **Purpose**: The purpose of the visualization is another essential factor to consider. Different techniques and tools better suit exploratory data analysis, presentation or reporting.
- **Interactivity**: Some data visualizations are designed to be highly interactive, while others are more focused on providing a clear, static representation of the data. The level of interactivity desired in the visualization will impact the choice of technique and tool.

- **Technical Skills**: The technical skills of the person creating the visualization will also impact the choice of technique and tool. Some tools may be more user-friendly and accessible, while others may require more advanced programming skills.
- **Integration**: If the visualization is to be integrated into a more extensive system or application, it is essential to consider the compatibility of the technique and tool with the existing technology stack.
- **Scalability**: Scalability is critical when future growth is expected as it affects how effectively the framework can manage larger datasets without sacrificing performance. The visualizations must remain responsive and usable despite size and complexity. This consideration has been found to be important by Palus, Poles and Posekany [52], who concluded that scalability should be a top factor to consider while choosing a visualization framework. Therefore, it is essential to examine the capacity of the selected framework and how it handles increased data volume to attest to its appropriateness.

Data visualization requires selecting the right method and tool for each situation. This selection should include data type, volume, purpose, interactivity, technical abilities and integration. By carefully evaluating these variables, organizations can choose the appropriate technique and tool for their data visualization needs and success.

5.5 THE FUTURE OUTLOOK FOR DATA VISUALIZATION

The future outlook for data visualization in enterprise intelligence is highly positive. As organizations collect and store ever-increasing amounts of data, the need for effective data visualization techniques and tools will only grow. Some several trends and developments are shaping the future of data visualization in enterprise intelligence:

- **Artificial Intelligence and Machine Learning**: Artificial intelligence and machine learning are increasingly important in data visualization. In the future, these technologies will automate the data visualization process, making it faster and more efficient.
- **Increased Interactivity:** As technology advances, the level of interactivity in data visualizations is expected to grow. This will enable organizations to explore and analyze their data in new and more powerful ways.
- **Virtual and Augmented Reality**: Virtual and augmented reality technologies also play an increasingly important role in data visualization. These technologies will enable organizations to experience their data more immersive and interactively.
- **Multi-Modal Visualization**: The trend towards multi-modal visualization, where data is visualized using multiple senses, is expected to continue. This will enable organizations to experience their data in new and more meaningful ways.
- **Big Data Analytics**: The continued growth of big data analytics drives the need for more effective data visualization techniques and tools. As big data analytics becomes

more widespread, the demand for data visualization tools that can handle large amounts of data is expected to increase.

Data visualization in enterprise intelligence has a bright future. Data visualization will continue to grow and evolve due to trends like artificial intelligence and machine learning, enhanced interaction, virtual and augmented reality, multi-modal visualization and big data analytics.

5.6 CONCLUSION

In conclusion, data visualization serves as a dynamic bridge between raw information and actionable insights. The utilization of visual tools like charts, graphs and maps empowers decision-makers within the realm of enterprise intelligence to unravel intricate data complexities, unveiling patterns and insights that guide well-informed choices. Through its capacity to present data with precision and conciseness, data visualization drives organizational communication and collaboration to new heights. Furthermore, the practical application of data visualization underscores its transformative impact across diverse industries and sectors. Frameworks provide structured pathways that assist professionals in crafting impactful visualizations. Looking forward, the horizon of data visualization holds promise, with technological advancements and analytical methods poised to amplify its capabilities. In an increasingly data-centric world, data visualization stands unwavering as a catalyst for informed decision-making and inventive problem-solving.

REFERENCES

1. Roberts, R., & Laramee, R. (2018). Visualising business data: A survey. In *Information* (Vol. 9, Issue 11, p. 285). MDPI AG. https://doi.org/10.3390/info9110285
2. Li, Q. (2020). Overview of data visualization. In *Embodying Data* (pp. 17–47). Springer Singapore. https://doi.org/10.1007/978-981-15-5069-0_2
3. Cavaller, V. (2021). Dimensional taxonomy of data visualization: A proposal from communication sciences tackling complexity. In *Frontiers in Research Metrics and Analytics* (Vol. 6). Frontiers Media SA. https://doi.org/10.3389/frma.2021.643533
4. Kwan, M.-P., & Lee, J. (2004). Geovisualization of human activity patterns using 3D GIS: A time-geographic approach. In D. Unwin & D. Hearnshaw (Eds.), *Visualization and Data Analysis 2004, Proceedings of SPIE* (pp. 84–95). SPIE.
5. Nothelfer, C., Lex, A., Strobelt, H., Streit, M., & Pfister, H. (2018). Visualization for predictive analytics in health care: A systematic literature review. *IEEE Transactions on Visualization and Computer Graphics*, 24(1), 424–434.
6. Liang, C., & Zhou, Y. (2020). Application of hierarchical clustering dendrogram in drug screening and analysis. *Journal of Molecular Graphics and Modelling*, 96, 1–11.
7. Liu, S., Li, Y., Li, J., Ma, K. L., & Plaisant, C. (2020). In-betweenness: A user-driven approach for visual exploration of time-series data. *IEEE Transactions on Visualization and Computer Graphics*, 27(6), 2890–2900.
8. Lee, J., Han, J., & Kang, S. (2019). Virtual and augmented reality for immersive data visualization. In *Handbook of Human-Computer Interaction* (pp. 1–16). Springer.

9. Li, R., Chen, X., Qin, Y., & Wang, H. (2020). Virtual reality and data visualization for exploring big data: Practices and challenges. *Engineering*, 6(5), 573–580.
10. Piumsomboon, T., Billinghurst, M., Kato, H., & Polvi, J. (2016). A systematic review of augmented reality visualizations for learning. *IEEE Transactions on Visualization and Computer Graphics*, 22(11), 2208–2224.
11. Muehlenhaus, I. (2014). *Visualization in the Age of Computerization*. Routledge.
12. Heer, J., Kong, N., & Agrawala, M. (2010). Sizing the horizon: The effects of chart size and layering on the graphical perception of time series visualizations. In *Proceedings of the SIGCHI Conference on Human Factors in Computing Systems* (pp. 1303–1312). Association for Computing Machinery.
13. Few, S. (2013). *Information Dashboard Design: Displaying Data for At-a-Glance Monitoring*. O'Reilly Media, Inc.
14. Shneiderman, B. (1996). The eyes have it: A task by data type taxonomy for information visualizations. In *Proceedings of the IEEE Symposium on Visual Languages* (pp. 336–343). IEEE.
15. Perer, A., & Shneiderman, B. (2008). Integrating statistics and visualization: Case studies of gaining clarity during exploratory data analysis. *ACM Transactions on Computer-Human Interaction (TOCHI)*, 15(2), 10.
16. Healy, K. (2018). *Data Visualization: A Practical Introduction*. Princeton University Press.
17. Satyanarayan, A., Moritz, D., Wongsuphasawat, K., & Heer, J. (2016). Vega-Lite: A grammar of interactive graphics. *IEEE Transactions on Visualization and Computer Graphics*, 23(1), 341–350.
18. Quinn, B. (2016). *Business Intelligence and Analytics Software Adoption Trends*. IDC Whitepaper.
19. Sood, T., & Agrawal, R. K. (2021). A review of enterprise dashboard design education models. *Journal of Information Systems Education*, 32(1), 63–72.
20. Wu, X., & Chen, S. (2017). Interactive visualization techniques in big data analytics: A review. *International Journal of Intelligent Systems*, 32(12), 1245–1272.
21. Fan, D., & Ma, X. (2019). Use of multiple visualization tools for enhanced data analysis and decision making. In *Proceedings of the 16th International Conference on Service Systems and Service Management (ICSSSM)* (pp. 1–5). IEEE.
22. Viana, S. (2017). Geovisualization and spatial data mining as decision support tools in retail location analysis. In *Advances in Data Mining. Applications and Theoretical Aspects* (pp. 205–217). Springer.
23. Pantazos, K., Asikis, T., Papadopoulos, S., & Vassilakopoulos, M. (2018). Network visualization and analysis for the industrial internet of things: A review. *IEEE Access*, 6, 10611–10622.
24. Mikroyannidis, A., Domingue, J., Phethean, C., & Beeston, G. (2018). Utilising learning analytics to identify pedagogical approaches to mobile inquiry-based learning. *British Journal of Educational Technology*, 49(6), 1044–1058.
25. Santos, K., Diedrichs, P., Seidler, N., & Cabrera-Umpierrez, M. (2019). Immersive analytics techniques for effective data visualization and analytics in industry 4.0. In *Proceedings of the 34th ACM/SIGAPP Symposium on Applied Computing* (pp. 2355–2362). Association for Computing Machinery.
26. Caballero-Ruiz, E., García-Sánchez, R., & González-Calderón, C. A. (2020). Data visualization techniques in healthcare. A systematic review. *Journal of Healthcare Engineering*, 2020.
27. Bailey, T. C., Chen, Y., Mao, L., & Lu, W. (2018). Applications of data visualization in public health: A scoping review. *Public Health Reviews*, 39(1), 1–28.
28. Peleg, M., Shahar, Y., & Quaglini, S. (Eds.). (2017). *Data Driven Healthcare: How Analytics and BI are Transforming the Industry* (Vol. 238). Springer.
29. Chargaff, M. (2019). The impact of visualization on stock market decisions. *International Journal of Advanced Research in Computer Science*, 10(4), 289–292.

30. Cheung, K. F., Lee, E. W., Chung, S. L., Lam, C. K., & Cheng, L. Y. (2018). Banking and finance applications in the cloud: The data visualization and visual analytics perspective. *Journal of Organizational Computing and Electronic Commerce*, 28(1), 73–92.
31. Li, Y., & Chen, Y. (2020). An IoT-enabled inventory management system for retail industry. In *2020 6th International Conference on Industrial Internet (ICII)* (pp. 51–55). IEEE.
32. Liu, L., Cai, H., Guo, J., Wang, J., Zhang, L., & Zhao, Y. (2018). Real-time traffic flow prediction based on big data from GPS-enabled smartphones. *Transportation Research Part C: Emerging Technologies*, 90, 150–164.
33. Hagy, S. (2018). Using data visualization to improve supply chain efficiency. *Transportation Journal*, 57(4), 383–395.
34. Yu, L., & Yang, C. (2019). Big data analytics in finance: A literature review. *Journal of Business Research*, 98, 97–108.
35. Lai, Y. (2017). The role of data visualization in financial analysis. *Journal of Financial Planning*, 30(10), 48–51.
36. Abi-Ghanem, R., Garcia, R., & de Almeida, V. T. (2020). AI-based neuromarketing: The next generation of marketing research. *Journal of International Technology and Information Management*, 29(2), 19–40.
37. Bajaj, A., & Coffey, R. (2019). Exploring the use and benefits of data visualization in business. *International Journal of Management, Information Technology and Engineering*, 7(8), 3111–3121.
38. Bostock, M., Ogievetsky, V., & Heer, J. (2011). D^3 data-driven documents. *IEEE Transactions on Visualization and Computer Graphics*, 17(12), 2301–2309. https://doi.org/10.1109/TVCG.2011.185.
39. Kulkarni, M., Dutta, A., & Ramesh, P. (2018). Dynamic data visualization using D3.js. In *2018 3rd International Conference on Computing Methodologies and Communication (ICCMC)* (pp. 265–270). IEEE.
40. https://public.tableau.com/s/
41. Power BI- Guided Learning Material. https://powerbi.microsoft.com/en-us/guided-learning/
42. Microsoft. (2019). Power BI overview. https://powerbi.microsoft.com/en-us/overview/
43. García, M., & Harmsen, B. (2012). *QlikView 11 for Developers*. Packt Publishing.
44. Harris, S. A., & Pedersen, C. S. (2018). QlikView. *Encyclopedia of Database Systems*, 2269–2274.
45. Wickham, H. (2016). *ggplot2: Elegant Graphics for Data Analysis*. Springer-Verlag. ISBN 978-3-319-24277-4; https://ggplot2.tidyverse.org
46. Wickham, H. (2016). ggplot2. *Wiley Interdisciplinary Reviews: Computational Statistics*, 3, 180–185. https://doi.org/10.1002/wics.1406
47. Noonpakdee, W., Khunkornsiri, T., Phothichai, A., & Danaisawat, K. (2018). A framework for analyzing and developing dashboard templates for small and medium enterprises. In *2018 5th International Conference on Industrial Engineering and Applications (ICIEA)*. IEEE. https://doi.org/10.1109/iea.2018.8387148
48. Zimmerman, C. J., Wessels, H. T. W. J., & Vatrapu, R. (2015). Building a social newsroom: Visual analytics for social business intelligence. In *2015 IEEE 19th International Enterprise Distributed Object Computing Workshop (EDOCW)*. IEEE. https://doi.org/10.1109/edocw.2015.20
49. Harrison, L., Reinecke, K., & Chang, R. (2015). Infographic aesthetics. In *Proceedings of the 33rd Annual ACM Conference on Human Factors in Computing Systems. CHI '15: CHI Conference on Human Factors in Computing Systems*. ACM. https://doi.org/10.1145/2702123.2702545
50. Mendoza, F., Paliwoda, A., & Bonardi, C. (2019). A framework for successful dashboard design. In *Proceedings of the 2019 International Conference on Information Systems* (pp. 1–11). Association for Information Systems.

51. Hullman, J., Resnick, P., & Adar, E. (2015). Hilbert's visions: Visualizing infinite data to inform research in mathematics education. In *Proceedings of the 2015 ACM Conference on Computer Supported Cooperative Work and Social Computing* (pp. 1701–1716).
52. Palus, S., Poles, R., & Posekany, A. (2021). Selection of the most appropriate visualization tools and techniques for big data analytics: Literature review. *Journal of Big Data*, 8(1), 1–24.

CHAPTER 6

Beyond the Hype

Understanding the Potential of ChatGPT in the Articulation of Technical Papers

Neha Shaah

6.1 INTRODUCTION

Technical writing is a crucial aspect of academic and scientific research, requiring a high level of accuracy, clarity and coherence to effectively communicate complex ideas. The process of writing a technical paper can be time-consuming and challenging, especially when dealing with large amounts of information. OpenAI has developed ChatGPT, a language generation model that can assist in the technical writing process. ChatGPT is a tool that is gaining quite a lot of attention lately.

The introduction of ChatGPT offers a solution to the difficulties faced in technical writing by providing a tool that can generate high-quality and coherent text. With the ability to handle large amounts of information and generate multiple perspectives and styles, ChatGPT can save time and effort while increasing the efficiency and accuracy of the writing process. GPT is now being used to develop virtual assistants that can help people with tasks like scheduling calls, sending emails, booking tickets, etc. [1].

This chapter explores the usefulness of ChatGPT in writing technical papers and covers the advantages and drawbacks of using the model, comparisons with other similar tools, and provides recommendations for effectively using ChatGPT to generate high-quality and informative technical writing. Understanding the potential of this tool and how to effectively use it is crucial in ensuring the accuracy and relevance of the generated text and improving the technical writing process. This being a general review where the overall scheme does not call for any implementation, and hence, the same is not covered in this chapter.

6.2 LITERATURE REVIEW

Large language models (LLMs) are typically trained on large set of text data, allowing them to accurately analyze and generate human-like text. These are trained using unsupervised learning – which means they are not explicitly provided with correct output for a given input, but instead they are expected to learn to generate reasonable outputs

DOI: 10.1201/9781032699806-6

based on given inputs. Analysts say that the NLP market is rapidly growing from $11B (2020) to $35B+ (2026) [2]. But this growth is not limited to market size alone. The model size and the number of parameters involved is also large. Wayne Xin Zhao et al. state in their study that when the number of parameters in these models exceeds a certain level, these enlarged language models not only achieve a significant performance improvement but also show some special abilities that are not present in small-scale language models [3].

There are several AI tools available that can assist with technical writing, including ChatGPT and its alternatives. ChatGPT is a large language model developed by OpenAI that can generate coherent and fluent text in a variety of styles and formats. It has been trained on a large amount of text data and can generate text that is similar to human-written text [5].

One alternative to ChatGPT is Google's Bard AI, which is powered by Google's in-house PaLM 2 large language model [6]. Another alternative is the new Bing chatbot, which runs on GPT-4, the newest version of OpenAI's language model systems [5]. This chatbot has access to the Internet and can provide information on current events, as well as link back to the sources it retrieved its answer from [5].

In terms of their use for technical writing, an experienced technical author explored using ChatGPT to assist with a number of writing projects and found that it can provide time-savings through drafts and prompting for additional content, but lacks accuracy and depth [7]. The author suggests working iteratively with ChatGPT, asking for small chunks of text with well-crafted prompts [7].

6.3 METHODOLOGY

It is important to understand how ChatGPT works in order to evaluate its accuracy or usefulness. ChatGPT is an interactive tool that accepts a prompt from user and provides a response [1]. It also comes with an ability to modify the given prompt and regenerate response with or without changing the prompt. This allows user to have an improvised version of the response or manually combine the various responses to suit their needs.

User needs to provide a specific input for what text they want to generate. User becomes an active part of this process as they will also have to assess the text generated and interactively get the LLM to generate text that meets user requirements with respect to tone of the text, accuracy, clarity on the topic and references. Figure 6.1 shows the workflow for execution of this technical paper writing task assigned to LLM. It also shows how user and LLM need to interact at various stages to get the best results.

6.4 COMPARISON WITH OTHER SIMILAR TOOLS

ChatGPT is not the only language generation model available. Other similar models include GPT-3, Grover and CTRL. These are advanced language generation models developed by

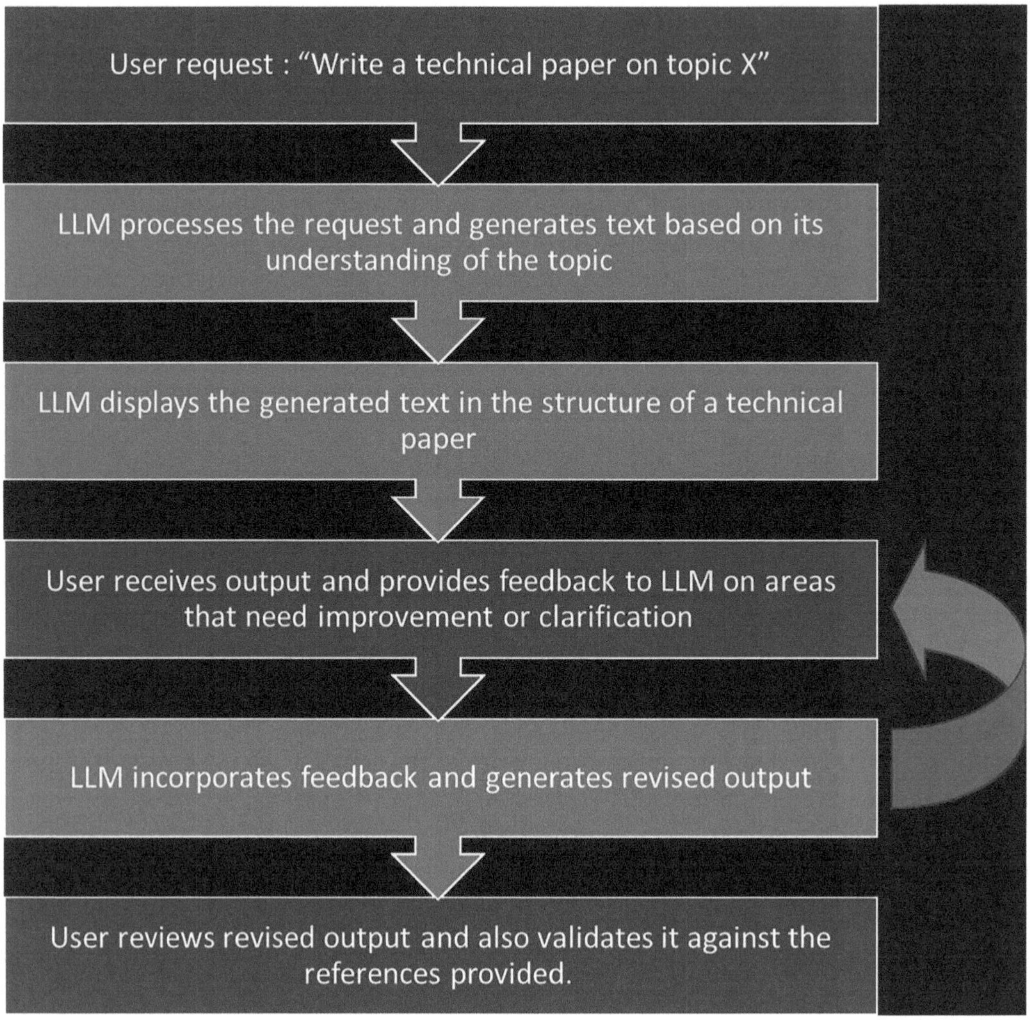

FIGURE 6.1 LLM workflow for technical paper writing.

OpenAI, which are considered to be better than ChatGPT in terms of their size, diversity and performance.

1. **Size**: GPT-3, Grover and CTRL are larger language models than ChatGPT, which allows them to handle a broader range of tasks and to generate more coherent and human-like text.
2. **Diversity**: GPT-3, Grover and CTRL have been trained on a much larger and diverse corpus of text than ChatGPT, which allows them to generate more varied and nuanced output.
3. **Advanced techniques**: GPT-3, Grover and CTRL use more advanced techniques, such as self-supervised learning, to improve the quality of their generated text.

4. **Fine-tuning:** GPT-3, Grover and CTRL can be fine-tuned for specific tasks, which allows them to generate text that is highly tailored to the needs of the user.
5. **Performance**: GPT-3, Grover and CTRL have been shown to outperform ChatGPT on a variety of language generation tasks, including language translation, question-answering and summarization.

There are other AI-powered language generation models similar to ChatGPT, including BERT and XLNet

- **BERT by Google**: BERT is a language representation model developed by Google that can be fine-tuned for language generation tasks. It offers high accuracy and is capable of handling complex tasks such as question-answering and sentiment analysis [6].
- **XLNet by Google**: XLNet is another language representation model developed by Google that outperforms BERT on certain NLP tasks and has the advantage of being permutation-based, allowing it to handle the dependencies between words in a sentence more effectively. Empirically, under comparable experiment settings, XLNet outperforms BERT on 20 tasks, often by a large margin, including question answering, natural language inference, sentiment analysis and document ranking [7].

Table 6.1 shows a comparison of various such models with ChatGPT. This table covers only high-level differences between various models. It is important to note that each of these models has its strengths and weaknesses, and the choice of which model to use

TABLE 6.1 Comparison of Various AI-Powered Language Generation Models

	Various AI-Powered Language-Generation Models					
Features	**BERT [8]**	**XLNet [9]**	**Grover**	**GPT3 [1]**	**CTRL [10]**	**ChatGPT [1]**
Pre-training	Unidirectional	Bidirectional	Unidirectional	Unidirectional	Conditional	Unidirectional
Pre-training Corpus	Wikipedia + BooksCorpus	Large-scale corpus	Webpages	Webpages + BooksCorpus	Webpages + BooksCorpus	Webpages
Task Performance	Strong in NLP tasks like QA, sentiment analysis, named entity recognition	Strong in NLP tasks, outperforms BERT in some cases	Strong in text generation tasks	Strong in text generation tasks, outperforms other models in some cases	Strong in text generation tasks with specific conditions	Good in text generation tasks
Fine-tuning	Can be fine-tuned for specific tasks	Can be fine-tuned for specific tasks	Can be fine-tuned for specific tasks	Can be fine-tuned for specific tasks	Can be fine-tuned for specific tasks with conditions	Can be fine-tuned for specific tasks
Model Size	Large (340M–375M parameters)	Largest (550M–600M parameters)	Moderate (1.5B parameters)	Largest (175B parameters)	Large (2B parameters)	Moderate (117M parameters)

depends on the specific requirements and needs of the user. Additionally, the performance of these models can vary depending on the task and data.

6.5 ADVANTAGES AND LIMITATIONS OF CHATGPT

The advantages of using ChatGPT for writing technical papers include its ability to generate high-quality, accurate and coherent text, its efficiency in handling large amounts of information, and its ability to generate multiple perspectives and styles. However, there are also some drawbacks to consider, such as its limitations in understanding context and making subjective judgments, as well as the potential for errors and inaccuracies in the generated text.

Evaluating the accuracy of text generated by ChatGPT requires careful attention and critical thinking. It is important to verify the information generated by ChatGPT against reliable sources to ensure that the information is accurate and up-to-date. Technical papers are highly contextual, and hence, relevance of generated text must be checked to match the context of its use, the purpose of paper and the intended audience. Although the generated text usually follows the semantics of good grammar, it is important to check the spellings and sentence structure to ensure that it is easily understood by the intended audience. And that the tone and style of the text is consistent throughout the paper – it does not contain and irrelevant or inconsistent information. The ultimate goal should be logically organizing the text so there is a smooth flow and it is easier to be followed.

Some limitations of ChatGPT with respect to writing a technical paper are listed in Table 6.1.

1. **Lack of domain expertise**: ChatGPT is trained on a vast amount of data, but it does not have the specialized knowledge or experience required to write a technical paper on a specific subject. It may generate text that lacks depth and accuracy in complex and specialized areas.
2. **Limited creativity**: Although ChatGPT can generate a wide range of styles and perspectives, it lacks the creativity and intuition required to write an engaging and original technical paper. The novelty of a technical paper should not be expected from text generated from pre-trained models.
3. **Inconsistent quality**: While ChatGPT can generate high-quality text, the quality of its output can be inconsistent, and it may produce text that is not well structured, poorly written, or lacks coherence.
4. **Ethical considerations**: Relying on ChatGPT to write a technical paper raises ethical concerns about the authorship and originality of the work, and it may be viewed as academic dishonesty.
5. **Inability to understand context**: ChatGPT operates based on the data it has been trained on and may not fully understand the context in which the technical paper is being written, leading to irrelevant or inaccurate information being generated.

6.6 RECOMMENDED PROCESS FOR USING CHATGPT IN CONTEXT OF TECHNICAL PAPERS

Having looked at the capabilities and limitations of ChatGPT and its applicability in case of technical writing, here are some guidelines on the process to be followed.

- **Dos:**
  1. Provide a clear prompt that defined the task for ChatGPT.
  2. Use specific keywords to ensure precise information.
  3. Proofread the output.
  4. Use multiple prompts and compile them together. Use one-shot or few-shot learning.
  5. Fine-tune the model to your specific domain and task to improve its performance and accuracy.
  6. Perform plagiarism checks
- **Don'ts:**
  1. Rely solely on ChatGPT: ChatGPT is a powerful tool, but it is not a substitute for human expertise. Always use your own judgement and knowledge to verify the accuracy and relevance of the generated text.
  2. Use it for unethical purposes – to generate text that is harmful or misleading or violates intellectual property rights.
  3. Ignore the limitations.
  4. Over-rely on a single prompt.
  5. Publish generated text without modification – always modify the generated text to ensure accuracy, clarity and coherence before publishing it.
  6. Trust the math or other calculations: It has been seen several times; it is easy to riddle ChatGPT and ask it to give a wrong answer, while it is also declared on their usage guidelines that the tool sometimes hallucinates and gives unexpected answers. Its best to use technical tools for testing your code, logic or math before putting it down in a technical paper.

6.7 ACCEPTANCE OF CHATGPT BY REPUTED JOURNALS OR PUBLICATIONS

The elephant in the room – "Does the work get accepted by reputed journals or conferences if I use ChatGPT to write it?"

Research integrity is already a pressing problem in academia [11] and websites such as RetractionWatch [12] convey a steady stream of fake, plagiarized and just plain wrong, research studies. Might ChatGPT make this problem even worse?

ChatGPT may not pass a plagiarism test as it is trained on a large corpus of text and may generate text that closely resembles existing sources. It is important to evaluate the generated text and ensure it is original and not copied from other sources, especially when submitting work for academic or professional purposes. Additionally, it is good practice to properly cite any information used from external sources, even if generated by a language model like ChatGPT.

This study has shown that ChatGPT can generate plausible-seeming research studies for well ranked journals. When private data and researcher expertise iterations are added to the research, the results would indeed be impressive. In my opinion, ChatGPT or any AI assistant should be looked at as an *e-research assistant* and hence become a new part-and-parcel tool of how research is normally carried out. This would, in a way, bring more equality across researchers across the globe, now that everyone has access to such assistance.

From ethical perspective, there are a couple of ways this could be addressed. ChatGPT-enabled research could be acknowledged on ethical research guidance frameworks, such as Elsevier's CRediT [13]. Iaia (2022) [14] suggests that higher levels of our generated research studies, with private data and iteration, could be considered the researcher's own work, but perhaps not the basic research study using only public data and simple question prompts.

> It ain't what you do, its the extent that you do it, and that's what gets (ethically-acceptable) results. [15]

A lot of agencies are actively discussing these topics and creating awareness on what is acceptable and what is not. The answer to the ethical issues is likely to be gradually understood rather than immediately apparent. One useful guide to how this might play out is how AI-generated work is treated under copyright laws of various countries.

6.8 CONCLUSION

Overall, bots like ChatGPT have the potential to be valuable tools for writers of technical documents. By providing suggestions for content and helping to improve readability, these models can help writers to create clear and effective technical documents. However, it is important to note that these models are not perfect and may require human oversight to ensure that the generated text is accurate and appropriate for the intended audience.

ACKNOWLEDGEMENT

I would like to acknowledge the use of OpenAI's language model, ChatGPT, as a tool to aid in writing process of this chapter. While ChatGPT was used to generate initial text, all final content was thoroughly reviewed, edited and revised by human author to ensure accuracy, originality and ethical compliance. As the main author who was responsible for conceptualization, structure and the linking of ideas, I feel that ChatGPT's contribution does not warrant co-authorship. However, I do acknowledge and appreciate the role that

ChatGPT played in the creation of this article, and I recognize the ethical and professional considerations involved in determining authorship.

The views and opinions expressed in this chapter are solely those of the author and do not represent the views or opinions of the author's organization. This research is a personal endeavor and is not affiliated with or endorsed by the author's organization. The findings and conclusions presented in this chapter are intended for academic and informational purposes only and should not be considered professional advice.

REFERENCES

1. ChatGPT. (n.d.). *Openai.com.* Retrieved May 20, 2023, from http://openai.com/product/chatgpt
2. The future of large language models (LLMs): Strategy, opportunities and challenges. (n.d.). *INDIAai.* Retrieved July 30, 2023, from https://indiaai.gov.in/article/the-future-of-large-language-models-llms-strategy-opportunities-and-challenges
3. W. X. Zhao, K. Zhou, J. Li, T. Tang, X. Wang, Y. Hou, Y. Min, B. Zhang, J. Zhang, Z. Dong, Y. Du, C. Yang, Y. Chen, Z. Chen, J. Jiang, R. Ren, Y. Li, X. Tang, Z. Liu, . . . J.-R. Wen. (2023). A survey of large language models. *arXiv [cs.CL].* http://arxiv.org/abs/2303.18223
4. Image source: Efficient natural language processing. (n.d.). *Mit.edu.* Retrieved July 30, 2023, from https://hanlab.mit.edu/projects/efficientnlp_old/
5. S. Ortiz. (2023, July 27). The best AI chatbots of 2023: ChatGPT and alternatives. *ZDNET.* https://www.zdnet.com/article/best-ai-chatbot/
6. M. Timothy. (2023, January 6). The 6 best alternatives to ChatGPT. *MUO.* https://www.makeuseof.com/best-alternatives-chatgpt/
7. M. Mason. (n.d.). Using ChatGPT as a technical writing assistant. *Martinfowler.com.* Retrieved July 30, 2023, from https://martinfowler.com/articles/2023-chatgpt-tech-writing.html
8. J. Devlin, M.-W. Chang, K. Lee, and K. Toutanova. (2018). BERT: Pre-training of deep bidirectional transformers for language understanding. *arXiv [cs.CL].*
9. Z. Yang, Z. Dai, Y. Yang, J. Carbonell, R. Salakhutdinov, and Q. V. Le. (2019). XLNet: Generalized autoregressive pretraining for language understanding. *arXiv [cs.CL].*
10. N. S. Keskar, B. McCann, L. R. Varshney, C. Xiong, and R. Socher. (2019). CTRL: A conditional transformer language model for controllable generation. *arXiv preprint arXiv:1909.05858.*
11. N. Mejlgaard et al. (2020). Research integrity: Nine ways to move from talk to walk. *Nature*, vol. 586, no. 7829, pp. 358–360.
12. F. Joelving et al. Retraction watch. *Retraction Watch.* [Online]. https://retractionwatch.com/ [Accessed: 20-May-2023]
13. Elsevier. CRediT author statement. *Elsevier.com.* [Online]. https://www.elsevier.com/authors/policies-and-guidelines/credit-author-statement [Accessed: 20-May-2023]
14. V. Iaia. (2022). To be, or not to be . . . original under copyright law, that is (one of) the main questions concerning AI-produced works. *GRUR International*, vol. 71, no. 9, pp. 793–812.
15. M. M. Dowling, and B. M. Lucey. (2023). ChatGPT for (finance) research: The Bananarama conjecture. *SSRN Electronic Journal*, vol. 53, p. 103662.

CHAPTER 7

Metaheuristics and Deep Learning in Lung Nodule Detection and Classification

Rama Vaibhav Kaulgud and Mandar Saundattikar

7.1 CLASSIFICATION MODELS FOR NODULE DETECTION

The basic steps of diagnosis are associated with the low-dose computed tomography (CT) imaging than the standard dose CT due to the survival rates, and low-dose CT supports the radiologists to detect the nodules. The examination of the patient from the CT images is a tedious task that seeks experienced knowledge and domain expertise, which duly affects the diagnosis accuracy [1]. Hence, the manual detection methods lack significance in highlighting the automated techniques, which would improve detection accuracy and earlier diagnosis. However, the heterogeneity of the nodules causes the models to yield unsatisfactory results. Traditionally, morphological techniques, threshold approaches, clustering using metaheuristics, and other AI-based models contribute much towards locating the nodules in the CT [2].

Morphological techniques are based on the non-linear operators concerning the nodule shape and texture of the nodules, and the morphology techniques follow the umbra function, while the threshold approaches depend on the pre-set threshold for segmenting and classifying the nodules [3]. Among all, metaheuristics and AI particularly, deep learning (DL) gains significance as the DL extracts and presents the fine-tuned features for classification, and there is no need for traditional textual descriptors, like local binary patterns (LBP), histogram of oriented gradients (HOG) and so on along with the other handcrafted features, like shape and texture of the nodules [4].

7.2 REVIEW OF THE CONVENTIONAL DEEP LEARNING MODELS

Though the review on the pulmonary nodules is not new, updated analysis always illustrates the current scenario in the research particularly, in deep learning and metaheuristics. Hence, this section reviews the traditional detection models based on deep learning and the necessity of metaheuristics is insisted. Figure 7.1 shows Stages of lung nodules. First of all, the advantages associated with the deep learning models over the traditional approaches are in the former's remarkable ability for data processing, robustness and ability

DOI: 10.1201/9781032699806-7

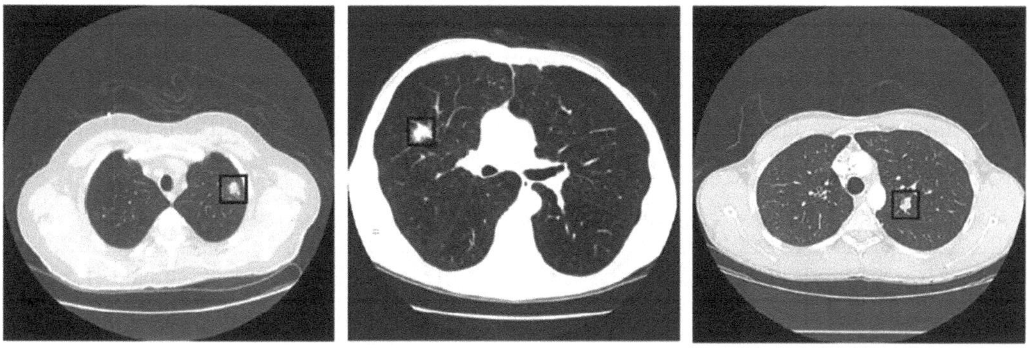

FIGURE 7.1 Stages of lung nodules – benign, primary and metastatic/malignant.

to tackle overfitting issues. Moreover, architecture-wise, the model is encapsulated with the pooling function that further minimizes the computational overhead, which along with the in-depth features of the CT image promotes classification accuracy. On the other hand, the training time contributes to the classification performance for which nowadays, metaheuristics has been used. Let us suppose the training sample is denoted as

$$T = \{X_i, Y_i\}, \text{ where } i = 1, 2, \ldots, n \tag{7.1}$$

where

X_i denotes the input CT and

Y_i is the label for the input sample.

7.3 LOSS FUNCTION FOR THE MODEL

Now, the model-based loss function is modelled as

$$L_1(P,\rho) = -\sum_{i=1}^{n} Y_i \log \mathrm{P}(Y_i = 1 | X_i, P, \rho) - \sum_{i=1}^{n} (1 - Y_i) \log \mathrm{P}(Y_i = 0 | X_i, P, \rho) \tag{7.2}$$

where, $L_1(P, p)$ is the loss function with P being the network parameters and ρ as the layer-wise network parameters. $\mathrm{P}(Y_i = 1 | X_i, P, \rho)$ and $\mathrm{P}(Y_i = 0 | X_i, P, \rho)$ are the probability of the ith sample with respect to the output class "0" and "1". Now, the choice for the classes depends on the fine-tuning of the network parameters, for which metaheuristics play a major role in selecting the best set of network parameters for the classifier so that the classification accuracy is improved

7.4 LOSS FUNCTION FOR THE OPTIMIZATION

The loss function-2 associated with the optimization performance is

$$L_2 = \sum_{i=1}^{n} M(O_i - O) \tag{7.3}$$

where, O_i is the acquired model output and O be the target, and L_2 should be minimal for the best set of the network parameters. The network parameters contributing to the minimal error function are chosen as the parameter set for the classifier and the loss functions of the model are explained and the review of the various training algorithms is detailed in the following.

7.5 METAHEURISTICS APPROACHES FOR DEEP LEARNING MODELS

In [5], Ying Su et al. clearly highlighted the significance of DL, and discussed the convolutional neural network (CNN) along with the different forms of CNN, where the feature generation module and the classification module are highlighted. Though the deep methods were capable of evaluating the features to support the classification problem, the capability to deal with the small objects is poor due to insufficient feature extraction. Coming to the network training, the weighted network is trained and the training speed decides the accuracy hence, the author highlights the need for the training algorithm. Furthermore, dilated units of CNN [6] for detection use the adaptive learning algorithm instead of ADAM optimizer for improving the learning performance. Moreover, in [7], a two-stage model is used that utilizes the ADAM optimizer for training and showing improved robustness. However, the problem was regarding the training scale that would affect the convergence rate, which prolonged the training time leading to computational complexity. On the other hand, 3D models have been discussed in [1], which outperform with improved parameter efficiency and model accuracy using the stochastic gradient descent approach (SGD). The drawback of the model was that the learning rate is the pre-defined one that seeks manual attention and expert knowledge.

Ratishchandra Huidrom et al. [8] employed a genetic algorithm (GA) and particle swarm optimization (PSO) for training the model, which aimed at managing the local optima convergence avoidance scenario. However, the learning ability of the algorithm was poor in training the deep models due to the parameter-sensitive settings. Similarly, in [9], PSO has been used for training CNN with the aim to decrease false positives. The optimization played well in enhancing the accuracy with minimal computational overload, but the point of global optimal convergence is the question. The recurrent-CNN model is highlighted in [10], graph-CNN in [11] and updated U-net in [12].

From these observations, there are some disadvantages, such as computational complexity, training time, efficiency, overlapping, convergence speed and data imbalanced problems. So this chapter uses lung nodule classification is used to overcome these drawbacks, as shown in Table 7.1, where the significance of deep learning and learning performance is demonstrated.

Figure 7.2 shows the various metaheuristics, and the role of these algorithms in learning the deep learning models is effective in enhancing the accuracy through minimizing the training time of the model. The genetic algorithm is based on the survival of the fittest and is best for the global searching abilities that have the tendency to faster convergence. Similarly, tabu, biogeography, forest and other optimizations listed under evolutionary intelligence ideas are known for their training time, while

TABLE 7.1 Deep Learning Models and Learning Algorithms

S. No.	Author	Method	Advantage	Disadvantage
1.	Ying Su et al. [5]	Faster R-CNN algorithm	Increasing the robustness of the network and reduced the probability of overfitting	The amount of data used in this experiment is too small, and it is old.
2.	Zeeshan Ali et al. [6]	End-to-end segmentation algorithm	Removed noise without blurring the edges of nodules	This method faced the data imbalance problem.
3.	Giovanni Lucca Franca da Silva et al. [9]	Convolutional neural network (CNN)	Reduced computational complexity	It contained an overfitting problem.
4.	Hongtao Xie et al. [13]	2D convolutional neural network (CNN)	Reduced the false positive candidates	The convergence speed is slow.
5.	Sunyi Zheng et al. [14]	Maximum intensity projection (MIP) based on CNN	Lower false-positive rate	It consumed more time in the training process.
6.	Xia Huang et al. [10]	Faster regional-CNN (R-CNN)	Reduced memory consumption and high processing speed	High overlapping problem.
7.	Marysia Winkels and Taco S. Cohen et al. [11]	3D CNNs with group convolutions (3D G-CNNs)	Easy implementation	It has a high computational cost.
8.	Chengwei Zhou et al. [12]	U-Net algorithm	Reduced the risk of missed diagnosis	It is more expensive and has low efficiency.

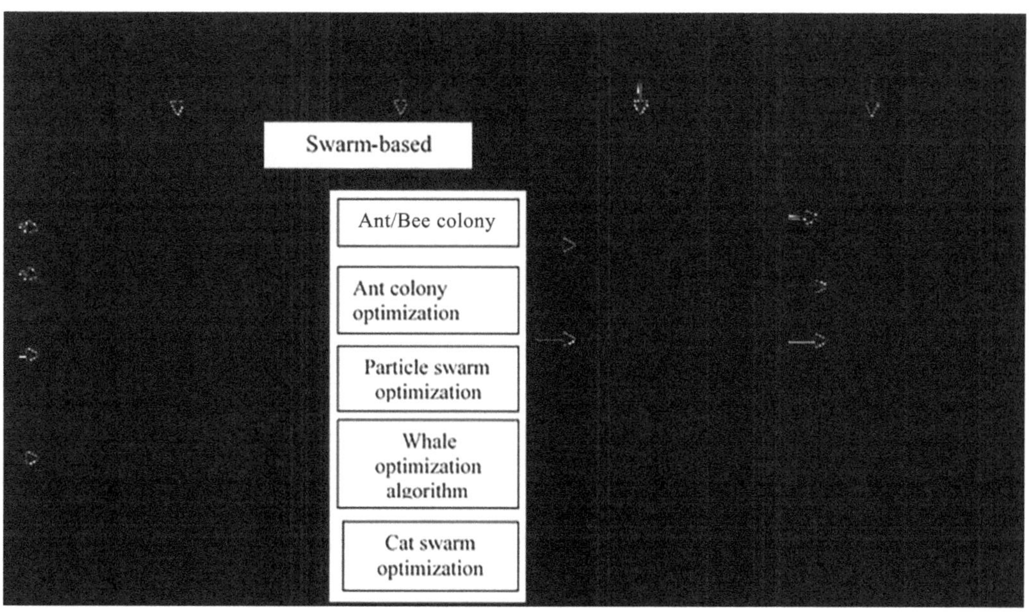

FIGURE 7.2 Various metaheuristics in model training.

the computation complexity associated with the training is large. Because there is no direct optimization algorithm, genetic algorithms can produce a wide range of outcomes, which is why they are so successful. A metaheuristic search technique used for mathematical optimization is called tabu search. By relaxing the local search's fundamental rule, tabu search improves its performance. Whenever the biogeography optimizer, which in turn can decrease needless intrusive testing and/or shorten diagnostic and therapeutic turnaround times. Another method for resolving non-linear optimization issues is the forest optimization algorithm (FOA). Human optimization is based on the searching and thinking abilities of the persons, while swarm intelligence is based on nature-inspired concepts.

The general challenges associated with deep metaheuristics are listed in the following. The challenges associated with the optimization are regarding the convergence rate that is responsible for the accurate tuning of the classifier. Furthermore, the scarcity of the data affects the classifier's performance and the sensitivity is insufficient. The improper training of the classifier causes huge computational complexity and affects the classification accuracy.

7.6 BASIC STEPS OF LUNG NODULE DETECTION

In this section, the stages of the detection are discussed in Figure 7.3. The CT image is acquired from the database and most of the literature presents the LIDC-IDRI dataset for nodule detection. Then, features are presented for segmentation and classification as cancerous and non-cancerous.

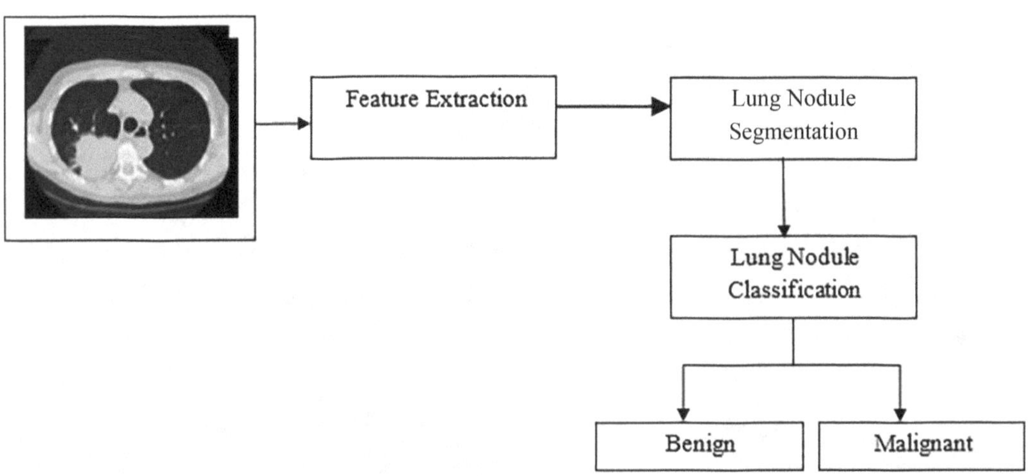

FIGURE 7.3 Basic stages of lung nodule detection and classification.

7.7 PERFORMANCE METRICS

The classification model is implemented and the analysis is undergone based on the standard metrics, such as accuracy, specificity and sensitivity. Table 7.2 shows Achievements of the Standard Metrics for the Models.

1. **Accuracy:** The accuracy represents the compactness of the specification, calculation and measurements that coordinates to a specific value. The accuracy is mathematically demonstrated as:

$$Acc = \frac{I_{tp} + I_{tn}}{I_{tp} + I_{tn} + I_{fp} + I_{fn}} \tag{7.4}$$

2. **Sensitivity:** Sensitivity is the percentage of the positive class that was correctly identified from the image:

$$sn = \frac{I_{tp}}{I_{tp} + I_{fn}} \tag{7.5}$$

3. **Specificity:** Sensitivity is the percentage of the negative class that was correctly identified from the image:

$$sp = \frac{I_{fn}}{I_{tn} + I_{fp}} \tag{7.6}$$

Where I_{tp} illustrates the true positive value and I_{tn} illustrates the true negative value, I_{fp} illustrates the false positive value, and I_{fn} illustrates the false negative value.

In the future, developing effective methods that ensure the effective balance between exploration and exploitation is the need, which promotes the proper balance between the intensification and diversification stages of the optimization. The optimization with the deep models should improve the classification accuracy by all means through the higher tendency towards the global optimal convergence.

TABLE 7.2 Achievements of the Standard Metrics for the Models

Measures	**Achievements (%)**
Accuracy	70–85
Sensitivity	Above 80 and below 83
Specificity	80–85

7.8 DATASET DESCRIPTION

The 1018 CT scan images from the LIDC-IDRI dataset [15], which was annotated by four renowned radiologists, were used for the experiments in this research. Only CTs with at least three annotations from experts are used. Cases classed as malignant have an average malignancy score of three or higher, whereas benign cases have an average malignancy score of three or lower. When cases with malignancy values of precisely 3 are removed, 1670 nodules are left behind. Each CT scan separates the slices next to a nodule from the other slices. The maximum area size associated with each nodule is 512 when the largest nodule depth size is reported to be 33 slices.

7.9 CONCLUSION

This chapter opens the necessity of deep models and the significance of metaheuristics in lung nodule detection and classification. Due to the body's rapid growth of cells, which is indicated by the appearance of pulmonary nodules, lung cancer is identified in the world health report as the second most common cancer in humans. The diagnosis of lung cancer involves the segmentation and classification of lung nodules, during which the pulmonary nodules are inspected for signs of cancer or an earlier stage. Through a higher inclination towards global optimal convergence, optimization with deep models should increase classification accuracy, sensitivity and specificity in all ways and also increase computational complexity. In the future research must focus on establishing efficient approaches that guarantee a healthy balance between exploration and exploitation to promote the intensification and diversification stages of optimization.

7.10 SUMMARY

The concept of lung nodule identification in particular has significant clinical importance and scientific value. Round or atypically shaped lesions called pulmonary nodules (PN) are common in the lungs. That typically appears as a dense shadow on lung computed tomographic (CT) images, with either clear or ambiguous boundaries. First and foremost, there are benefits to deep learning models over conventional methods in terms of their outstanding data processing capabilities, resilience and capacity to handle overfitting difficulties. Along with the intricate features of the CT image, the model's architecture-wise encapsulation of the pooling function further reduces the computational cost and enhances classification accuracy.

REFERENCES

1. Nasrullah, Nasrullah, Sang, Jun, Alam, Mohammad S., Mateen, Muhammad, Cai, Bin, and Hu, Haibo. "Automated lung nodule detection and classification using deep learning combined with multiple strategies," *Sensors (Basel)*, vol. 19, no. 17, p. 3722, 2019.
2. Cao, Haichao, Liu, Hong, Song, Enmin, Ma, Guangzhi, Xu, Xiangyang, Jin, Renchao, Liu, Tengying, and Hung, Chih-Cheng. "Two-stage convolutional neural network architecture for lung nodule detection," *IEEE Journal of Biomedical and Health Informatics*, vol. 24, no. 7, pp. 2006–2015, 2020.

3. Halder, Amitava, Chatterjee, Saptarshi, Dey, Debangshu, Kole, Surajit, and Munshi, Sugata. "An adaptive morphology based segmentation technique for lung nodule detection in thoracic CT image," *Computer Methods and Programs in Biomedicine*, vol. 197, p. 105720, 2020.
4. Zuo, Wangxia, Zhou, Fuqiang, Li, Zuoxin, and Wang, Lin. "Multi-resolution CNN and knowledge transfer for candidate classification in lung nodule detection," *IEEE*, vol. 7, pp. 32510–32521, 2019.
5. Su, Ying, Li, Dan, and Chen, Xiaodong, "Lung nodule detection based on faster R-CNN framework," *Computer Methods and Programs in Biomedicine*, vol. 200, p. 105866, 2021.
6. Ali, Zeeshan, Irtaza, Aun, and Maqsood, Muazzam, "An efficient U-Net framework for lung nodule detection using densely connected dilated convolutions," *The Journal of Super Computing*, vol. 78, no. 2, pp 1602–1623, 2022.
7. Cao, Haichao, Liu, Hong, Song, Enmin, Ma, Guangzhi, Xu, Xiangyang, Jin, Renchao, Liu, Tengying, and Hung, Chih-Cheng. "A two-stage convolutional neural networks for lung nodule detection," *IEEE Journal of Biomedical and Health Informatics*, vol. 24, no. 7, pp. 2006–2015, 2020.
8. Huidrom, R., Chanu, Y. J., and Singh, K. M. "Pulmonary nodule detection on computed tomography using neuro-evolutionary scheme," *Signal, Image and Video Processing*, vol. 13, no. 1, pp. 53–60, 2019.
9. da Silva, G. L. F., Valente, T. L. A., Silva, A. C., de Paiva, A. C., and Gattass, M., "Convolutional neural network-based PSO for lung nodule false positive reduction on CT images," *Computer Methods and Programs in Biomedicine*, vol. 162, pp. 109–118, 2018.
10. Huang, X., Sun, W., Tseng, T. L. B., Li, C., and Qian, W. "Fast and fully-automated detection and segmentation of pulmonary nodules in thoracic CT scans using deep convolutional neural networks," *Computerized Medical Imaging and Graphics*, vol. 74, pp. 25–36, 2019.
11. Winkels, M., and Cohen, T. S., "Pulmonary nodule detection in CT scans with equivariant CNNs," *Medical Image Analysis*, vol. 55, pp. 15–26, 2019.
12. Zhou, Chengwei, Zhao, Xiaodong, Zhao, Lili, Liu, Jiayuan, Chen, Zixuan, and Fang, Shuai. "Deep learning-based CT imaging in the diagnosis of treatment effect of pulmonary nodules and radiofrequency ablation," *Computational Intelligence and Neuroscience*, vol. 2022, 2022.
13. Xie, H., Yang, D., Sun, N., Chen, Z., and Zhang, Y. "Automated pulmonary nodule detection in CT images using deep convolutional neural networks," *Pattern Recognition*, vol. 85, pp. 109–119, 2019.
14. Zheng, S., Guo, J., Cui, X., Veldhuis, R. N., Oudkerk, M., and Van Ooijen, P. M., "Automatic pulmonary nodule detection in CT scans using convolutional neural networks based on maximum intensity projection," *IEEE Transactions on Medical Imaging*, vol. 39, no. 3, pp. 797–805, 2019.
15. Dataset. https://www.kaggle.com/datasets/zhangweiled/lidcidri.

CHAPTER 8

An Improved Face Recognition Method Using Canonical Correlation Analysis

Ganesh D. Jadhav, Suhas Patil, Bhushan M. Borhade and Yogesh Shinde

8.1 INTRODUCTION

In day-to-day life, the rate of crime is high. So it has to require more security and public safety against all kinds of crime. One of the more often used biometric recognition technologies for safety and security measurement is the face recognition system [1–3]. So it is fully effective against all kinds of crime. Most of the country uses face recognition in its face recognition technology for security and safety of the public [4, 5]. Many advances in security research have been made, and many more are still being conducted. The approaches that are simpler, use less memory and handle data in real time are the emphasis of this research. Another is the feature fusion methods, which are also employed as a training set to enhance the system's effectiveness, accuracy and efficiency. In feature fusion analysis, the recognition process combines data from two or more feature vectors into a single vector, making the algorithms' accuracy and efficiency more effective. This technique combines data from several stages of the recognition system, although the research only using feature level.

8.2 OVERVIEW OF LITERATURE SURVEY

The literature on face recognition systems is divided into two parts. It is the most important biometric recognition system, in which face detection is one component and face identification is another [3]. In addition, the study uses feature fusion technique to improve the accuracy of the system as well as the efficiency of the algorithm.

8.3 FACE DETECTION

The study starts by talking about the face detection procedure. The initial step was to identify the face image using a face detection technique, such as the local binary pattern technique [6]. For testing purposes, the study employed both coloured and greyscale photos in proposed system. The study entirely committed to using the local binary pattern technique

 DOI: 10.1201/9781032699806-8

in this system. For image textures, the local binary pattern technique is most frequently employed. In this method, the histogram is produced by segmenting the image of a human face into discrete areas. Three sections make up the facial detection component [6].

1. Colour segmentation
2. Post-processing
3. Template

The ensuing algorithms for detection and recognition are determined by face representation, which is a model of a face [1, 5]. These methods are employed to calculate how similar two photos are. It is crucial for the security and safety of the general populace. Because a single person's face image might vary in stance, illumination, age, facial expression and makeup, it also presents some general computer vision issues. Consequently, it is a research area of interest with a variety of properties. In this project, it also applied the feature fusion technique to speed up computation and improve algorithm accuracy.

8.4 FACE IDENTIFICATION

The fisherface and the eigenface methods are two crucial methods for face identification [7]. There are two sections to the genuine image data collection. The needed testing photos are in the first half of the dataset, while the training images are in the second. Both the images are colourful and greyscale. The eigenface and the fisherface approaches are then used to calculate the outcome [3]. The eigenface method used in these techniques is a dimensional reduction technique known as the principal component analysis algorithm (PCA algorithm). The work employed linear discriminant analysis (LDA) and principal component analysis (PCA) to build the projection matrix for the fisherface approach [3, 8]. As a result, the algorithm's final equation matches the test images.

8.5 CANONICAL CORRELATION ANALYSIS: FUSION FOR MULTIMODAL BIOMETRIC RECOGNITION

However, in this study, the system solely used the fusion of feature levels. In the feature fusion technique, recognition is done at the combination or decision level. Additionally known as multimodal recognition [9]. While being uncorrelated inside each feature set, CCA-based feature fusion exploits the correlation between two sets of features to find two sets of transformations across the two feature sets [10]. Therefore, it is the most crucial method for raising the algorithm's precision and effectiveness.

8.6 GAP ANALYSIS IN EXISTING SYSTEM

1. For the face recognition system, Euclidean distance and a perfect match between training and test photos are used to determine the accuracy of the system. However, only the Euclidean Distance between averages of images was considered in this study.
2. The study may use feature fusion utilizing canonical correlation analysis to enhance this Face detection.

3. In the current investigation, it used the Yale face database, which has 165 greyscale photographs of 15 participants. In the future, the study will employ datasets comprising 210 images of 35 subjects.

4. The goal of the feature Fusion technique is to combine the data from two or more feature vectors into a single vector for recognition; however, in this system, it has only merged the data from two feature vectors into a single vector. These are employing the feature fusion method based on canonical correlation to combine this data.

8.7 METHODOLOGY

1. **Database:**
   a. 210 images out of 35: it can also add new images to the database.
   b. The Yale face database has 160 greyscale pictures with 15 people in them.
2. **Software Language:** as it is implemented in MATLAB®
3. **System Architecture:** There are two components to the facial recognition system. Face detection and face identification are both a part of this system, which is the most significant biometric recognition system. Additionally, the system applies the feature fusion technique to enhance both the algorithm's effectiveness and the system's correctness. Figure 8.1 shows Proposed face recognition system and Figure 8.2 shows Data flow for the same.

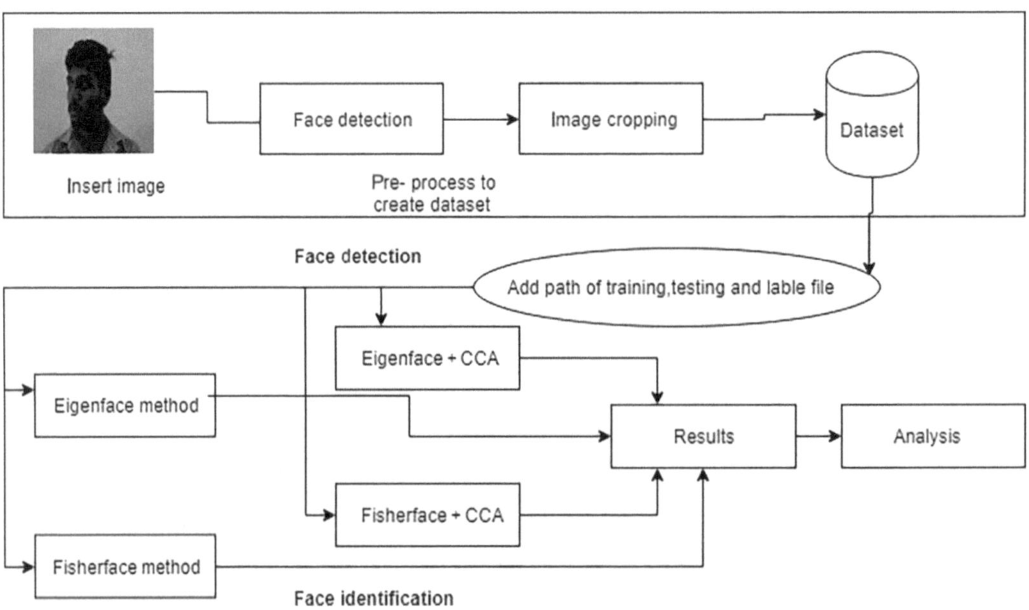

FIGURE 8.1 Proposed face recognition system.

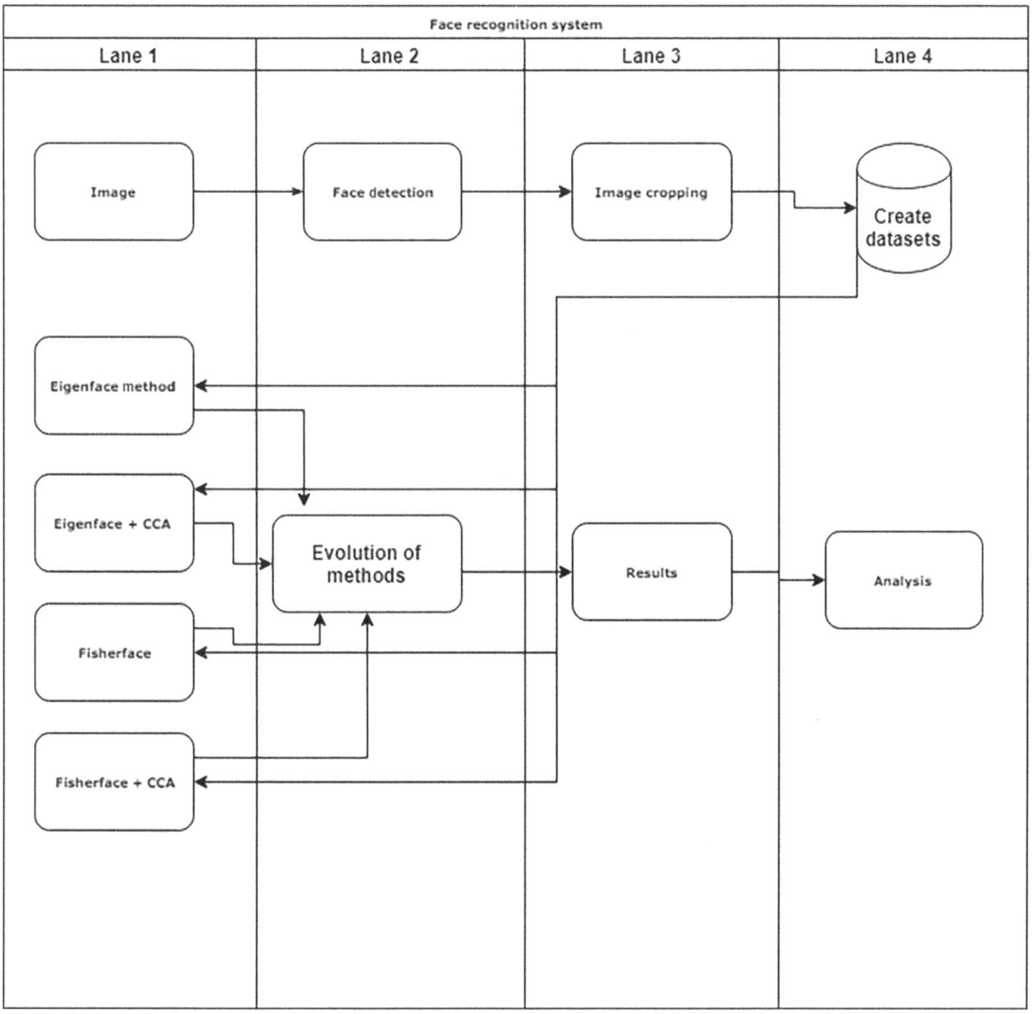

FIGURE 8.2 Data flow of face recognition system.

4. **Face Detection:** The study will start by talking about the face detection procedure. The initial step was to identify the face image using a face detection technique, such as the local binary pattern technique [4, 6]. For testing purposes, it employed both coloured and greyscale photos in our system. The study entirely committed to using the local binary pattern technique in this system [4, 6]. For image textures, the local binary pattern technique is most frequently employed [4, 11].

   In this method, the image of a human face is separated into region to generate the histogram. Figure 8.3 shows original LBP operator.

5. **Image Cropping:** After face detection in face recognition, it can proceed to image cropping using the MATLAB function "imcrop", an improved face recognition

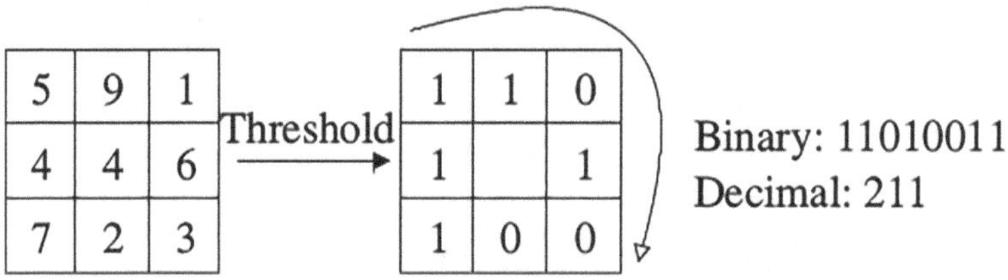

FIGURE 8.3 The original LBP operator.

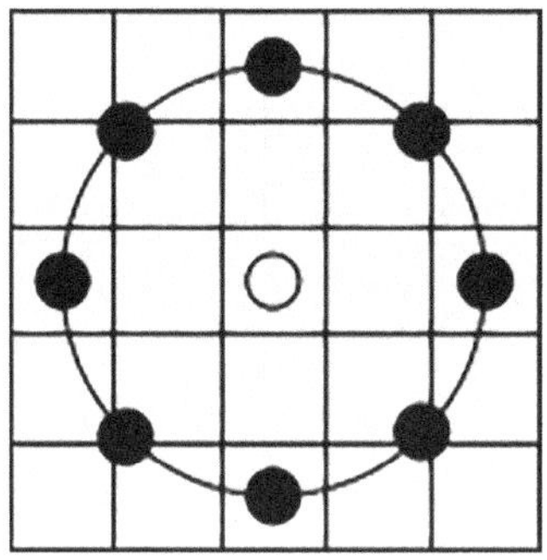

FIGURE 8.4 Circularly neighbour-set.

method using local binary pattern method [4]. This function allows us to build the dataset by cutting the face section into rectangles. Both training and testing data are included in the data collection. Figure 8.4 shows Circularly neighbour-set and image cropping is shown in Figure 8.5.

6. **Face Identification:** The fisherface and the eigenface methods are two crucial methods for face identification [4, 7]. There are two sections to the genuine image data collection. The needed testing photos are in the first half of the dataset, while the training images are in the second. Both the images are colourful and greyscale. Histogram of cropped image is shown in Figure 8.6 and Figure 8.7 shows Face detection.

   The eigenface and the fisherface methods are then used to calculate the outcome [7]. The eigenface technique used in this method is a dimensional reduction technique known as the principal component analysis algorithm (PCA algorithm). The work employed linear discriminant analysis (LDA) and principal component analysis (PCA) to build the projection matrix for the Fisherface approach [4].

7. **Eigenfaces:** This algorithm needed the following fundamental pre-processes:

   The system must build the training dataset and the testing dataset in order to set up the database in its fundamental configuration. It needed 270 photos in total for 45 real-time datasets in order to identify the faces. And there are 160 greyscale pictures of 15 people in the Yale face database.

   The next step after building the database is to calculate the eigenfaces and fisherfaces. The study retains the M images from the training dataset for the eigenface

FIGURE 8.5 Image cropping.

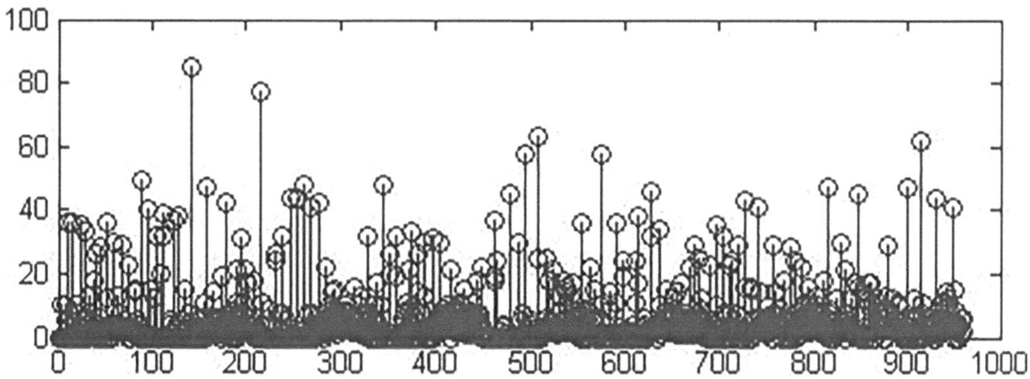

FIGURE 8.6 Histogram of cropped image.

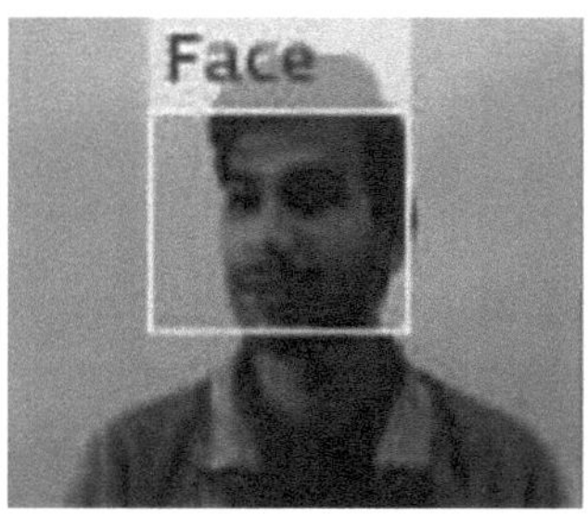

FIGURE 8.7 Face detection.

method. M is hence the eigenvalue of the specific dataset. The photos' dataset is created by the face space. The data collection is automatically updated when new photographs are discovered in it.

In the face recognition system, the study processes faces through the eigenface method. So the study intends to transition towards the training sets that are created through the use of N images. The training set is i (i = 1, 2, . . . N). So 100 × 100 is the dimension of each image. The matrix is converted through the training image dataset.

$$A=[\Phi 1, \Phi 2, \ldots \Phi n] \tag{8.1}$$

Finding a fresh set of features that are linear combinations of the initial ones is what we're after.

In the above set Φi is an image representation of column matrix [4].

- $\Phi i=\phi - \mu$
- ϕ=reshape(Γi,[mn, 1])
- $\mu = meani(\phi i)$

Then it has to calculate the scatter matrix. The scatter matrix is as follows:

$$S_T = AA^T \tag{8.2}$$

After the calculation of the scatter matrix of the training dataset, the study moves towards the linear transformation to determine the W of the eigenvectors of the training dataset. When the eigenvector is determined, the important factor is the feature space, which will be p-dimensional. p = N mn will be created based on the observation of the condition. In the calculation of the scatter matrix of an image, to maximize the determinant of W, PCA chooses W [4].

$$\mathrm{W}_{opt} = argmax|W^T S_T \mathrm{W}| = [w_1, w_2, w_3, \ldots, w_p] \tag{8.3}$$

After the calculation of the scatter matrix of the image, the eigenvectors are Wi. So the largest eigenvalue is p. When the

The scatter matrix is calculated, and then the dimension of the feature face is automatically reduced up to p. At the conclusion of the neural distance calculation, the testing images reveal their closest match. Figure 8.8 shows Algorithm of eigenface + CCA.

8. **Fisherfaces:** This algorithm required some basic pre-processing per the eigenface algorithm. It has completely focused on our algorithm. The fisherface technique is the same as the eigenface, but it has some advantages. Within the class and between the class and the total distribution of the training set are the advantages of the fisherface method.

   After the pre-processes at the initialization of the fisherface method, the following are done: Firstly, the study required to set the path of the dataset into the algorithm. In the training set, this method employs multiple face images for each class. In order

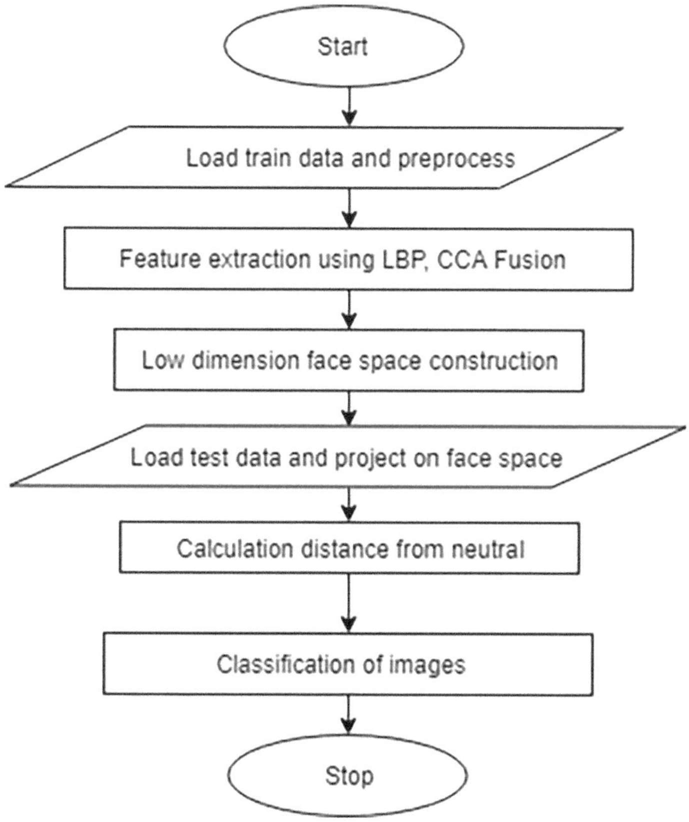

FIGURE 8.8 Algorithm of eigenface + CCA.

to proceed, it must first establish the training set, which is done by suggesting i photos from M subjects.

The training set is = $\{\Gamma_1, \Gamma_2, \Gamma_3, \Gamma_4, \ldots\ldots,_M\}$

In the equation, the set of image matrices Γi contains the training images [4]. It's a set of face images. The total images are classified into 35 classes, with each class having the image of only one person. So the number of classes is defined as follows:

$$X_1 = \{\Gamma_1, \Gamma_2, \Gamma_3, \Gamma_4, \ldots_M\}$$

$$X_2 = \{\Gamma_5, \Gamma_6, \Gamma_7, \Gamma_8\}$$

$$\ldots\ldots\ldots\ldots$$

$$Xn = \{\Gamma_{m-3}, \Gamma_{m-2}, \Gamma_{m-1}, \Gamma_m)$$

For the advantages of the feature face algorithm, here is a three-scatter matrix. The scatter matrix for the within-class (Sw), between-class (Sb), and total distribution (St) these scatter matrices are for fisherface data images.

$$S_t = \Sigma^{M}{}_{n=1}(\Gamma n - \Psi)(\Gamma n - \Psi)^{T}\ldots\ldots\ldots\ldots \tag{8.4}$$

The average image is important to improve the efficiency of the algorithm. In the fisherface, it's required for the first average.

The image training dataset is $\Psi = \Sigma^{M}{}_{n=1}\, \Gamma n$. The second is the average image of each class, which is

$$\Psi i = \underline{1}\, \Sigma\, \Gamma$$

$$|Xi|\; \Gamma i \epsilon Xi$$

In the fisherface process, to create the projection matrix Upca, projection matrix is generated through principle component analysis (PCA) on the scatter matrix using dominant M-C components [4]. To reduce dimensionality within class, it can use the projection matrix Upca. The cl is the eigenvector of the scatter matrices, and Ufld is as follows:

$$U_{fld} = \boldsymbol{argmax} \frac{\left|U^{T}U^{T}_{pca}S_{B}U_{pca}U\right|}{\left|U^{T}U^{T}_{pca}S_{w}U_{pca}U\right|}$$

At the last step of the fisherface algorithms, the projection matrix Uff is calculated. Finally, the face images on the c matrix are maximized if the scatter between classes is minimized and within scatter is minimized for each class. So the final equation is defined as follows [4]:

$$Uff = UfldUpca$$

The same equation used in the eigenface to identification of the faces in face recognition system.

9. **Canonical Correlation Analysis (CCA):** In the feature fusion technique, the recognition is to combine the information from two or more feature vectors into a single one, but in this algorithm, it used pairwise correlation within the class. The fusion of information can occur at different levels of the recognition system, like feature level, matching score level, or decision level. It is also called multimodal recognition. CCA-based feature fusion uses the correlation between two sets of features to find two sets of transformations across the two feature sets while being uncorrelated within each feature set [10].

Projection matrix Upca. The c 1 is the eigenvectors of the scatter matrices, and U_{fld} is as follows [4].

Per the algorithm of CCA, there is a mutual relationship between two sets of variables. Here are two matrices: are XϵRp×n and Y ϵRq×n and each of them contains n training feature vectors from two different modalities. So, for each sample, two feature vectors are extracted from the first and second modalities. The vectors have p and q dimensions.

Per the previous figure, within the set covariance matrices are SxxϵRp×p and SyyϵRq×q of X and Y. Also the between-set dimension in this algorithm finally observed that the scatter [4]:

$$S = \begin{pmatrix} cov(x) & cov(x,y) \\ cov(y,x) & cov(y) \end{pmatrix} = \begin{pmatrix} S_{xx} & S_{xy} \\ S_{yx} & S_{yy} \end{pmatrix}$$

CCA aims to find the linear combinations, $X^* = W_x^T X$ and $Y^* = W^T Y_y$, that maximize the pair-wise correlations across the two feature sets [4]:

$$corr(\overset{*}{X},\overset{*}{Y}) = \frac{cov(\overset{*}{X},\overset{*}{Y})}{var(\overset{*}{X}).var(\overset{*}{Y})}$$

where $cov(X^*, Y^*) = W^T S$ W, $var(X^*) = W^T S$ W

and var(Y) = W_y SyyWy. The transformation matrices, Wx and Wy, are then found by solving the eigenvalue equations [10]:

Where Wx and Wy are the eigenvectors and R2 is the diagonal matrix of eigenvalues or squares of the canonical correlations.

As defined in [10], fusion is performed by concatenation or summation of the transformed feature vectors; it depends upon the

$$\begin{cases} S_{xx}^{-1} S_{xy} S_{yy}^{-1} S_{yx} \hat{W}_x = R^2 \hat{W}_x \\ S_{yy}^{-1} S_{yx} S_{xx}^{-1} S_{xy} \hat{W}_y = R^2 \hat{W}_y \end{cases}$$

mode of the string compare function. But in this project, the mode of the string compare function is "sum".

$$Z = \overset{*}{X} + \overset{*}{Y} + W_x^T X + W_y^T Y = \begin{pmatrix} W_x \\ W_y \end{pmatrix}^T \begin{pmatrix} X \\ Y \end{pmatrix}$$

This method is useful to improve the accuracy of the eigenface and the fisherface algorithms. Algorithm of CCA is shown in Figure 8.9.

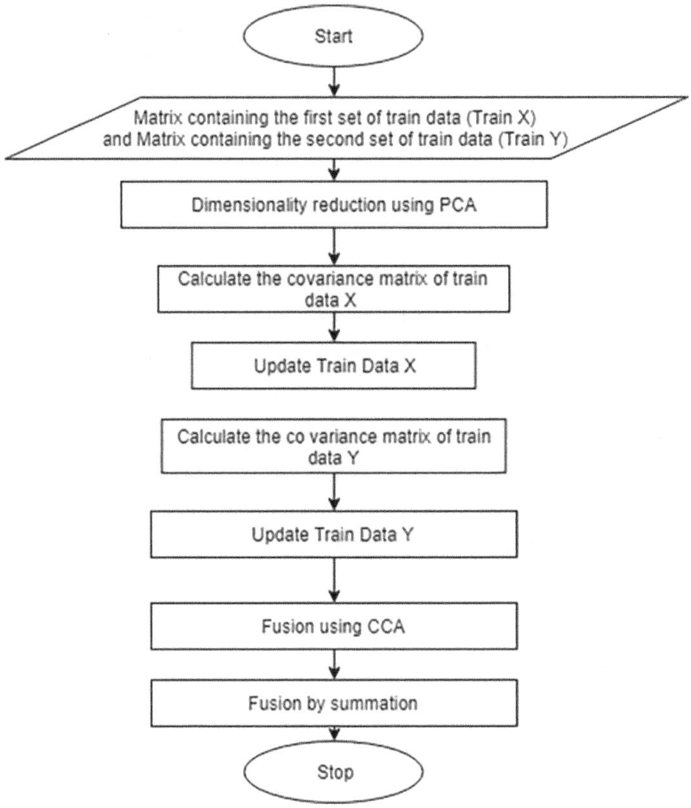

FIGURE 8.9 Algorithm of CCA

8.8 RELATED WORK

This literature focuses on the face recognition system using canonical correlation analysis in which the process is separated into two parts. The first part is face detection, which uses local binary pattern, and the second part is face identification [4]. In face identification, it has eigenface and fisherface algorithms, along with the feature fusion method by canonical correlation analysis.

1. **Face Detection:** Face recognition methodology: The first part needed to detect the face image by using the face detection method, like the local binary pattern method. In our system, it used greyscale as well as coloured images for the testing process [4]. In this system, the study completely focused on the local binary pattern method. After the face detection, next step moves towards the input image cropping.

2. **Image Cropping:** In face recognition, after the face detection, it can move towards image cropping, which is done using a MATLAB function "imcrop". Using this function, it can cut the face portion into a rectangular shape to create the dataset [4].

3. **Face Identification:** After the face detection, the study comes to the face identification, where the study completely focused on the fisherface algorithm.

   The last step of working is to find the better performance of the system using both methods along with the feature fusion technique using canonical correlation analysis. Accuracy and time consumption will be calculated using MATLAB code on the database [4].

4. **Feature Fusion Techniques Based on Canonical Correlation Analysis (CCA):** In the feature fusion technique, recognition is accomplished by combining data from two or more feature vectors into a single vector; however, in this system, it has simply merged the data from two feature vectors into a single vector [10]. The study employs the feature fusion method based on canonical correlation analysis to combine this data [10]. Our main focus is on using watermarking to transmit photographs securely, wherein only the authorized user is able to receive the undistorted photos. It takes the photograph and watermark it at the user's request. The study delivers it to the user so they may examine the image and check the watermark [8].

8.9 RESULTS AND DISCUSSION

In system, the study puts into words the results. In this system, the real image dataset is divided into two parts: the first part is the training dataset, and the second is the testing dataset. The database has 35 classes, and each class has six different images. Result of matching test images to train images is shown in Figure 8.10.

In face recognition system, both algorithms used the same datasets for all processes in identification.

Whenever the process is started, it wants to require the first image in one class as a testing image and the other images as training images. So it has to create testing datasets that contain the first image in each class and training datasets that contain all other face images.

The work calculates the matching testing image with train images using the eigenface method per Figure 8.10 and puts all that data into a graph. It has followed the same process for all other methods. At the first calculation of matching images with the test images using eigenface method set the path of training image dataset and testing image dataset. When the process will be completed, then calculate the accuracy of particular algorithm

Depending on this calculation, put all the values in the graph per Figure 8.10. At the first calculation of matching images with the test images using the fisherface method, set the path of the training image dataset and the testing image dataset. When the process is completed, calculate the accuracy of the particular algorithm. Depending on this calculation, put all the values in the graph per Figure 8.11.

At the first calculation of matching images with the test images using the eigenface + CCA method, set the path of the training image dataset and the testing image dataset. When

Accuracy of Eigenface

Training images per class	Accuracy of Eigenface
Training images-2	85.3658
Training images-3	85.3658
Training images-4	90.2439
Training images-5	95.1219

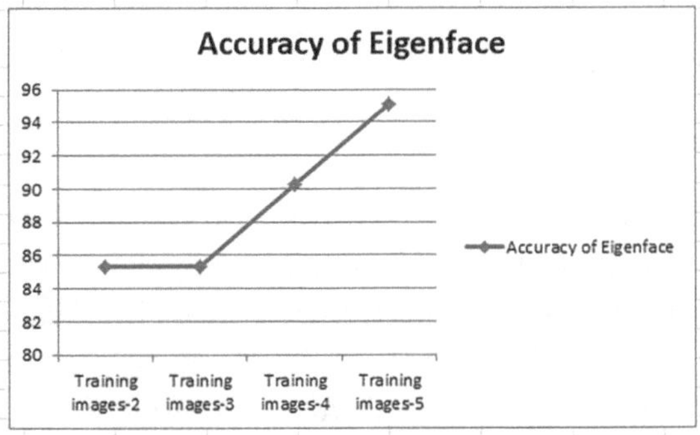

FIGURE 8.10 Accuracy for the eigenface method.

the process is completed, calculate the accuracy of the particular algorithm. Depending on this calculation, put all the values in the graph per Figure 8.12.

From the observation of Figure 8.13, it is shown that how accuracy in increasing the eigenface method using the CCA method.

```
//Test Image, Distance From Neutral, Match Person ID, Best Match
1.jpg,164,D01,D012.jpg
10.jpg,331,D10,D104.jpg
11.jpg,269,D11,D116.jpg
12.jpg,312,D12,D126.jpg
13.jpg,290,D13,D136.jpg
14.jpg,307,D14,D145.jpg
15.jpg,329,D15,D152.jpg
16.jpg,386,D16,D162.jpg
17.jpg,307,D17,D176.jpg
18.jpg,370,D18,D184.jpg
19.jpg,319,D19,D194.jpg
2.jpg,273,D02,D026.jpg
20.jpg,359,D20,D202.jpg
21.jpg,354,D21,D216.jpg
22.jpg,298,D13,D135.jpg
23.jpg,298,D23,D236.jpg
24.jpg,307,D24,D242.jpg
25.jpg,337,D25,D252.jpg
26.jpg,303,D26,D266.jpg
27.jpg,311,D27,D273.jpg
```

FIGURE 8.11 Result of matching test images to train images.

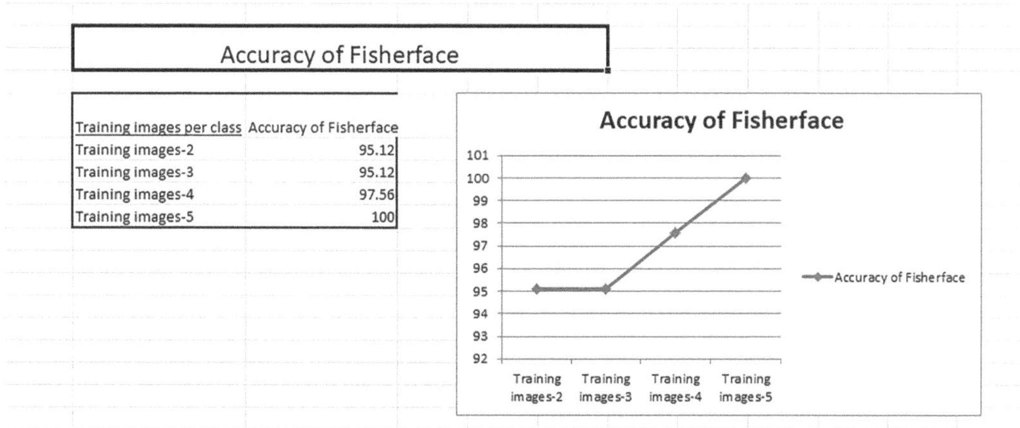

Training images per class	Accuracy of Fisherface
Training images-2	95.12
Training images-3	95.12
Training images-4	97.56
Training images-5	100

FIGURE 8.12 Accuracy for the fisherface method.

At the first calculation of matching images with the test images using the fisherface + CCA method, set the path of the training image dataset and the testing image dataset. When the process is completed, calculate the accuracy of the particular algorithm. Depending on this calculation, put all the values in the graph per Figure 8.13.

From the observation of Figure 8.14, it is shown that how the accuracy increases using the fisherface method using CCA. Figure 8.15 shows accuracy for fisherface vs. fisherface + CCA method.

Figure 8.16 shows how canonical correlation analysis can be used to improve the efficiency and accuracy of eigenface and fisherface face recognition systems. Eigenface method efficiency improves by 2.44%, and fisherface accuracy is 100% with only three images in the training dataset.

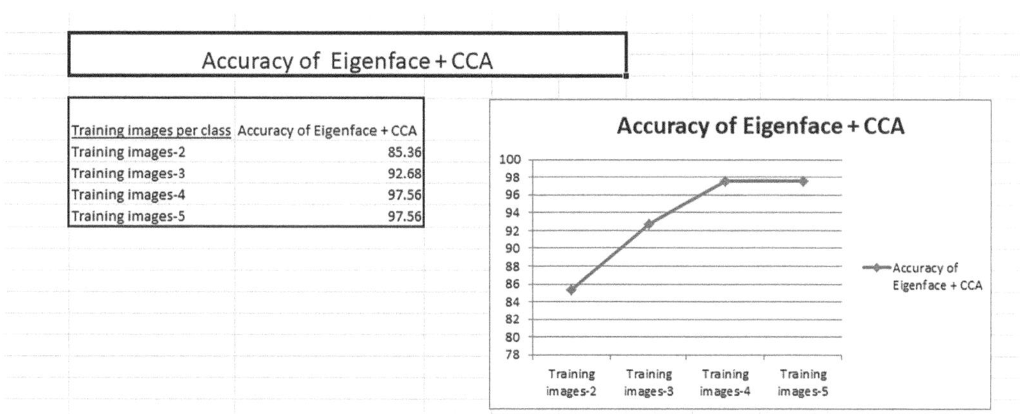

Training images per class	Accuracy of Eigenface + CCA
Training images-2	85.36
Training images-3	92.68
Training images-4	97.56
Training images-5	97.56

FIGURE 8.13 Accuracy for eigenface + CCA method.

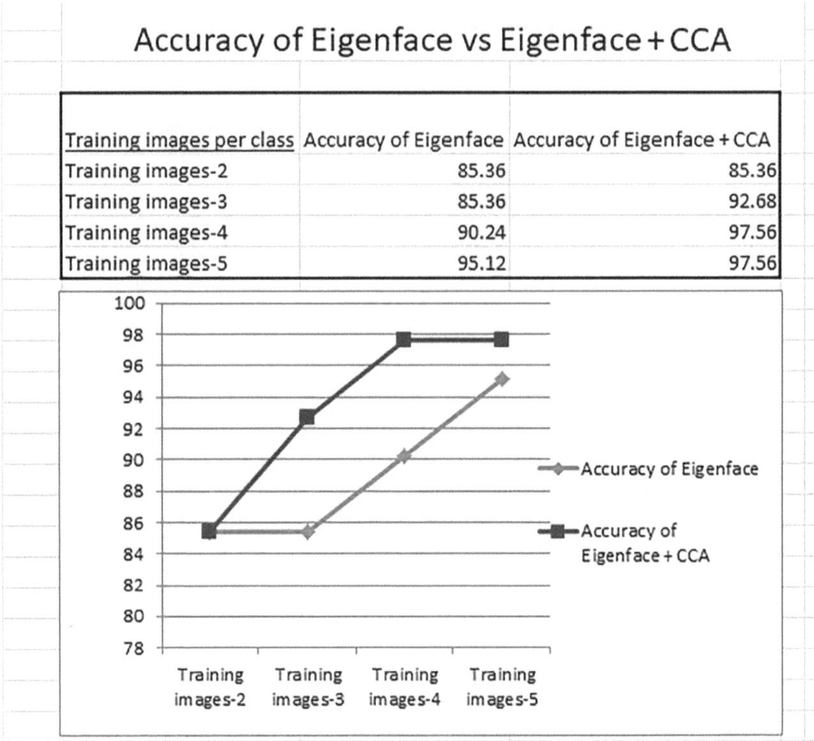

Accuracy of Eigenface vs Eigenface + CCA

Training images per class	Accuracy of Eigenface	Accuracy of Eigenface + CCA
Training images-2	85.36	85.36
Training images-3	85.36	92.68
Training images-4	90.24	97.56
Training images-5	95.12	97.56

FIGURE 8.14 Accuracy for eigenface vs. eigenface + CCA.

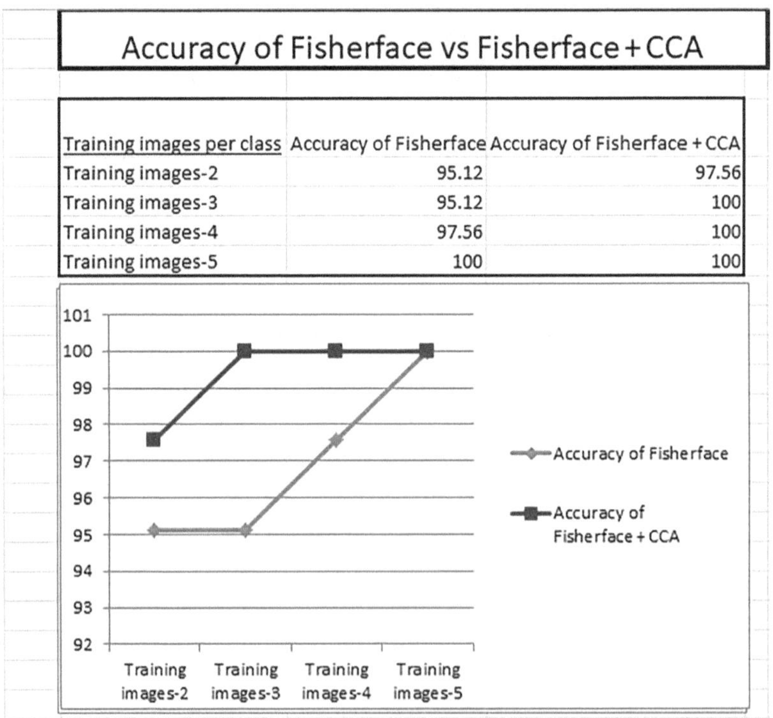

Accuracy of Fisherface vs Fisherface + CCA

Training images per class	Accuracy of Fisherface	Accuracy of Fisherface + CCA
Training images-2	95.12	97.56
Training images-3	95.12	100
Training images-4	97.56	100
Training images-5	100	100

FIGURE 8.15 Accuracy for fisherface vs. fisherface + CCA method.

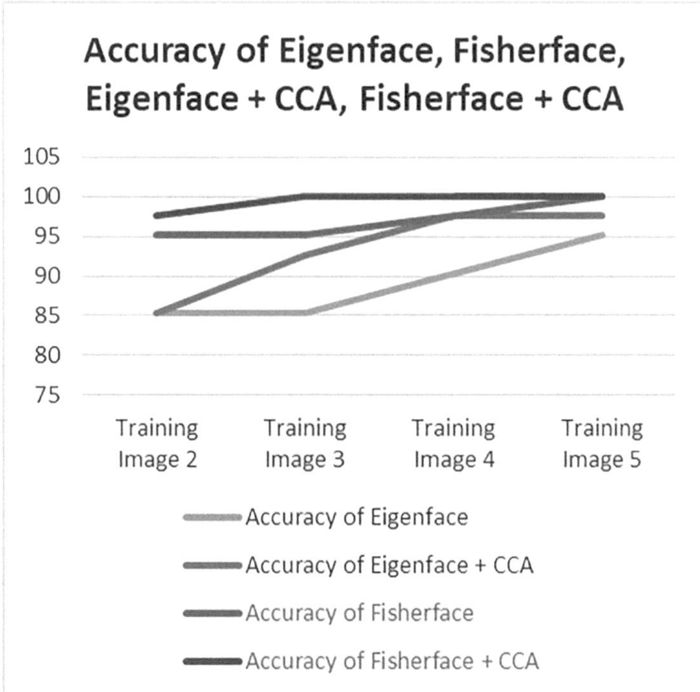

FIGURE 8.16 Final analysis.

8.10 CONCLUSION

Our goal is to improve the efficiency and accuracy of face recognition systems using the eigenface and fisherface methods by using feature fusion based on canonical correlation analysis. Eigenface efficiency improvement: The text mentions a 2.44% improvement in efficiency for the eigenface method. This could refer to faster computation or reduced computational resources required for recognition, making the method more efficient. Fisherface accuracy: The text claims a 100% accuracy rate for the fisherface method with only three images in the training dataset. This is a remarkable achievement and suggests that the fisherface method, in combination with the proposed feature fusion approach, can achieve perfect recognition results even with a small amount of training dataset.

The main objective is to enhance the performance of face recognition systems by utilizing the eigenface and fisherface methods and fusing their features through canonical correlation analysis. The reported improvements in efficiency and accuracy indicate promising advancements in the field of face recognition technology.

REFERENCES

1. Y.-H. Lee, H. Ahn, H.-J. Cho, and J.-H. Lee, "Advanced Face Recognition and Verification in Mobile Platforms," *Journal of Systems and Information Technology*, vol. 16, no. 2, pp. 126–137, 2014.

2. Xiaofei He, Shuicheng Yan, Yuxiao Hu, P. Niyogi, and Hong-Jiang Zhang, "Face Recognition Using Laplacianfaces," *IEEE Transactions on Pattern Analysis and Machine Intelligence*, vol. 27, no. 3, pp. 328–340, March 2005.
3. S. Z. Li, and Juwei Lu, "Face Recognition Using the Nearest Feature Line Method," *IEEE Transactions on Neural Networks*, vol. 10, no. 2, pp. 439–443, March 1999.
4. S. A. Saleh, S. Azam, K. C. Yeo, B. Shanmugam, and K. Kannoorpatti, "An Improved Face Recognition Method Using Local Binary Pattern Method," In *2017 11th International Conference on Intelligent Systems and Control (ISCO)*, Coimbatore, India, pp. 112–118, 2017.
5. M. Turk, and A. Pentland, "Eigenfaces for Recognition," *Journal of Cognitive Neuroscience*, vol. 3, no. 1, pp. 71–86, 1991.
6. M. A. Rahim, M. N. Hossain, T. Wahid, and M. S. Azam, "Face Recognition Using Local Binary Patterns (LBP)," *Global Journal of Computer Science and Technology Graphics Vision*, vol. 13, no. 4, pp. 1–7, 2013.
7. P. N. Belhumeur, J. P. Hespanha, and D. J. Kriegman, "Eigenfaces vs. Fisherfaces: Recognition Using Class Specific Linear Projection," *IEEE Transactions on Pattern Analysis and Machine Intelligence*, vol. 19, no. 7, pp. 711–720, July 1997.
8. X. Tang, and X. Wang, "Face Photo Recognition Using Sketch," In *Proceedings of International Conference on Image Processing* (Vol. 1, pp. I–I). IEEE, 2002.
9. K. I. Chang, K. W. Bowyer, and P. J. Flynn, "An Evaluation of Multimodal 2D+3D Face Biometrics," *IEEE Transactions on Pattern Analysis and Machine Intelligence*, vol. 27, no. 4, pp. 619–624, April 2005.
10. Mohammad Haghighat, and Wadee Alhalabi, "Discriminant Correlation Analysis: Real-Time Feature Level Fusion for Multimodal Biometric Recognition," *IEEE Transactions on Information Forensics and Security*, vol. 11, no. 9, September 2016.
11. Zehang Sun, G. Bebis, and R. Miller, "On-road Vehicle Detection: A Review," *IEEE Transactions on Pattern Analysis and Machine Intelligence*, vol. 28, no. 5, pp. 694–711, May 2006.

CHAPTER 9

Guesswork to Results

How ML-Based A/B Testing Is Changing the Game

Namita K. Shinde, Payal Kadam, Aditya Choudhary, Bhavay Chopra and Krishnansh Awasthi

9.1 INTRODUCTION

Digital marketing has revolutionized the way businesses reach their target audience and drive sales. With the increasing use of digital marketing tools such as Google Analytics, Google AdWords and Facebook Ads, A/B testing has become a popular method used by businesses to assess the impact of their digital marketing initiatives [1]. A/B testing allows businesses to determine the effectiveness of slight modifications made to landing pages, popup forms, article titles and other digital marketing decisions [2]. However, traditional statistical methods used in A/B testing have limitations, which fail to capture the complexity of customer behaviour. This chapter explores the use of machine learning (ML) predictive models in A/B testing to overcome these limitations and provide valuable insights for decision-making [3]. In recent years, businesses have turned to A/B testing to measure the effectiveness of their digital marketing campaigns [4]. A/B testing involves creating two versions of a website or marketing campaign, with slight variations in design or messaging, and then measuring the performance of each version [5]. This process helps businesses determine which version of their campaign is more effective in driving customer engagement and, ultimately, sales [6]. Traditional statistical methods used in A/B testing rely on hypothesis testing and confidence intervals to determine whether the results of an experiment are statistically significant. These methods have been widely used in A/B testing and are considered the gold standard in statistical analysis. However, traditional statistical methods have several limitations, including the inability to capture complex customer behaviour and the lack of predictive power [7]. A subfield of artificial intelligence, machine learning (ML) allows computers to extract information from data and progressively improve their performance. ML algorithms can be used to create predictive models that can analyse large datasets and identify patterns and trends that are not immediately apparent through traditional statistical analysis [8]. In recent years, businesses have started

DOI: 10.1201/9781032699806-9

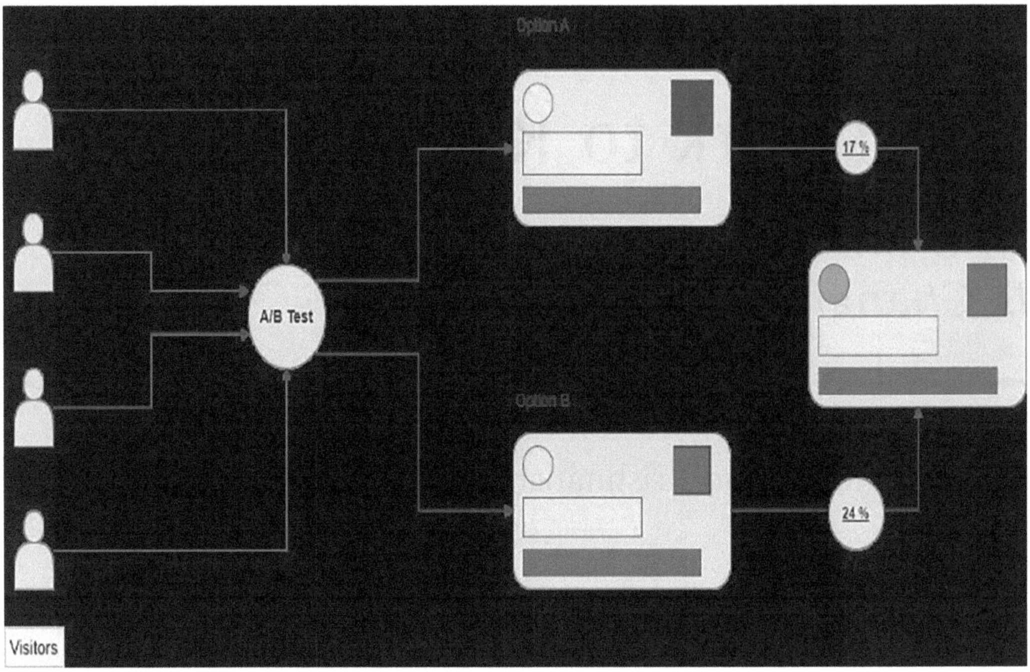

FIGURE 9.1 Visualizing AB testing.

to use ML algorithms to improve the accuracy and effectiveness of their digital marketing campaigns. Visualization of AB testing. is shown in Figure 9.1.

ML algorithms have several advantages over traditional statistical methods in A/B testing. ML algorithms can capture the complexity of customer behaviour and identify patterns that are not immediately apparent through traditional statistical analysis [9] Additionally, ML algorithms can create predictive models that can be used to optimize digital marketing campaigns and improve decision-making [10]. The purpose of this chapter is to explore the use of ML predictive models in A/B testing and demonstrate their effectiveness in improving the accuracy and effectiveness of digital marketing campaigns. This chapter presents a case study demonstrating the effectiveness of the ML-based tool in improving A/B testing results [11].

9.2 RELATED WORK

Several studies have been conducted on the use of ML algorithms in A/B testing. These studies have shown that ML models can improve the accuracy of A/B testing by identifying patterns in customer behaviour that are not immediately apparent through traditional statistical analysis. For example, a study conducted by Kohavi et al. showed that ML algorithms could improve the accuracy of A/B testing by up to 30% [12]. Other studies have also shown the effectiveness of ML algorithms in improving A/B testing results. A study conducted by Agarwal et al. showed that ML algorithms could identify hidden patterns in customer behaviour and provide insights that were not immediately apparent through traditional statistical analysis. Similarly, a study conducted by Tong and Zhang (2010) showed that ML algorithms could improve the accuracy of A/B testing by up to 10% [13].

Machine learning is a rapidly evolving field that has transformed numerous industries with its ability to make accurate predictions and uncover insights from large datasets. Linear regression, decision trees and XGBoost are three popular machine learning models that have proven to be effective in various applications. In this literature review, we will discuss how these models are being used in different domains and their strengths and limitations [14]. Linear regression is a simple yet powerful machine learning model that is widely used for predicting numerical values it functions by identifying the line that best fits the data and minimizes the disparity between the actual and anticipated values [15]. In the field of finance, linear regression is used to predict stock prices and currency exchange rates. In healthcare, it is used to predict patient outcomes and disease progression [16]. However, although there may not always be a linear relationship between both the dependent and independent variables linear regression maintains one, hold true in complex datasets [17]. Another well-liked machine learning model that is applied to both classification and regression applications is decision trees [18]. They operate by breaking down the data into more manageable groupings based on the most important characteristics, until the subsets contain homogeneous data [19]. Decision trees are used in various domains, such as finance, healthcare and marketing. In finance, they are used to predict credit risk and fraud detection [20]. In healthcare, they are used to predict patient outcomes and diagnosis [21]. However, decision trees can overfit the data, which may result in poor generalization and prediction accuracy [22].

XGBoost is a relatively new machine-learning model that has gained popularity due to its high accuracy and speed [23]. Several decision trees are combined in this ensemble classification algorithm to get reliable predictions. XGBoost has been used in various domains, such as finance, healthcare and computer vision [24]. In finance, it is used for portfolio optimization and fraud detection. In healthcare, it is used for predicting disease outcomes and diagnosis [25]. XGBoost has been shown to outperform other machine learning models in many applications, but it requires a large amount of data and careful parameter tuning [26]. In conclusion, machine learning models like linear regression, decision trees and XGBoost have proven to be effective in various domains. Linear regression is simple and easy to interpret but may not capture complex relationships in the data [26]. Decision trees can be prone to overfitting but are effective in handling categorical data [27]. XGBoost is a powerful and accurate model but requires careful tuning and a large amount of data [28]. These models have their strengths and limitations, as well as the model selection, should be based on the nature of the issue at hand and the properties of the data [29].

9.3 METHODOLOGY

9.3.1 Data Selection

Data selection is an essential step in developing an effective A/B testing model using machine learning. The data selected for the model should be representative of the target audience and provide relevant insights into customer behaviour [30]. The first step in data selection is to define the target audience and the specific digital marketing initiative to be tested. The target audience should be clearly defined based on demographic, geographic, psychographic and behavioural characteristics. The digital marketing initiative to be tested

```
START -> Collect Data -> Pre-process Data
Pre-process Data -> Split Data -> Training Data
Split Data -> Testing Data
Training Data -> Train Linear Regression Model -> Model Trained?
Model Trained? -> Yes -> Predict Test Data -> Evaluate Model -> Model Improved?
Model Improved? -> Yes -> Train Decision Tree Model -> Model Trained?
Model Trained? -> Yes -> Predict Test Data -> Evaluate Model -> Model Improved?
Model Improved? -> Yes -> Train XGBoost Model -> Model Trained?
Model Trained? -> Yes -> Predict Test Data -> Evaluate Model -> Model Improved?
Model Improved? -> No -> END
Model Improved? -> No -> Train Linear Regression Model with Different Parameters -> Model Trained?
Model Trained? -> Yes -> Predict Test Data -> Evaluate Model -> Model Improved?
Model Improved? -> No -> Train Decision Tree Model with Different Parameters -> Model Trained?
Model Trained? -> Yes -> Predict Test Data -> Evaluate Model -> Model Improved?
Model Improved? -> No -> Train XGBoost Model with Different Parameters -> Model Trained?
Model Trained? -> Yes -> Predict Test Data -> Evaluate Model -> Model Improved?
Model Improved? -> No -> END
```

FIGURE 9.2 Flowchart of the ML algorithm.

should also be defined, along with the specific goals and metrics to be measured [31]. Once the target audience and digital marketing initiative have been defined, data can be collected from various sources, including website analytics, customer relationship management (CRM) systems and social media platforms. The data collected should be relevant to the digital marketing initiative being tested and should include both the control and experimental groups [32]. The data collected should be cleaned and pre-processed to remove any missing values, outliers, or irrelevant data. Data pre-processing involves transforming the data into a format that is suitable for machine learning algorithms. This can include scaling, normalization and feature extraction [33]. The data selected should be representative of the target audience and provide valuable insights into customer behaviour. The quality of the data selected is essential in developing an accurate and effective A/B testing model using machine learning. [34]. Figure 9.2 shows the flow of the given ML algorithm.

9.3.2 Model Selection

The evaluation of the algorithms involved measuring their performance based on metrics, such as accuracy, precision and recall. The metrics were chosen to evaluate how well the model could predict customer behaviour in A/B testing [35].[1] Accuracy was used to measure the overall performance of the model, while precision was used to measure how often the model was correct when it predicted customer behaviour would change [36].[2] The recall was used to measure how often the model correctly identified when customer behaviour changed [37].[3] The algorithms evaluated included linear regression, logistic regression, decision trees, random forests and gradient boosting. These algorithms were chosen based on their proven effectiveness in other fields of machine learning and data analysis. Each algorithm was evaluated using a combination of pre-processed data and synthetic data to test their ability to predict customer behaviour in A/B testing [38].[4]

9.3.3 Model Training

The selected ML algorithm was trained on the pre-processed data. The model was optimized using hyper parameter tuning to improve its predictive accuracy [39]. Hyper parameter tuning involved fine-tuning the model's parameters to optimize its performance. Splitting the data into training and validation sets allowed the model to be trained on the training set and verified on the validation set during the training process [40]. A training set and a validation set were created from the data. The model was trained using the training set, and its effectiveness was evaluated using the validation set. Gradient descent was used to train the model, which involved incrementally changing the model's parameters to reduce the discrepancy between the expected and actual results [41]. Model was improved after each iteration of the training process, which was repeated numerous times. Based on how well it performed on the validation set, the final model was selected [42]. Figure 9.3 shows visual representation of the flowchart of the proposed A/B testing tool.

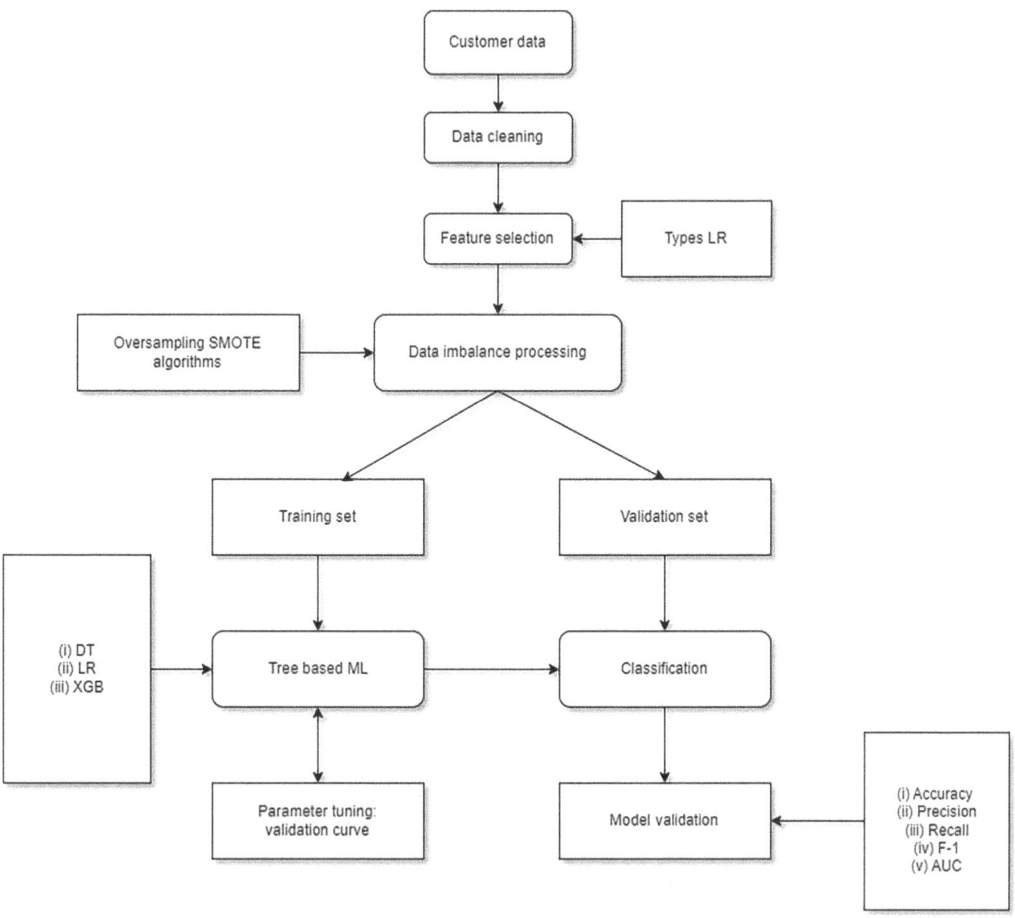

FIGURE 9.3 Visual representation of the flowchart of the proposed A/B testing tool.

9.3.4 Model Evaluation

The trained model was evaluated on a separate test set of data to determine its performance. The evaluation was based on various metrics, including accuracy, precision and recall. The results of the evaluation showed that the model could accurately predict customer behaviour in A/B testing [43]. The evaluation involved comparing the predicted output of the model to the actual output. The model's performance was measured using various metrics, including accuracy, precision and recall. The results of the evaluation were used to fine-tune the model's parameters to optimize its performance [44].

9.3.5 Tool Development

The ML-based tool was developed using Python and various libraries, including sci-kit-learn, pandas and Flask [45]. The tool was designed to take in A/B testing data and output predictions of customer behaviour. The tool's user interface was developed using Flask, and it was designed to be easy to use and user-friendly [46]. The tool's user interface was designed to be intuitive and user-friendly, allowing users to easily input data and view the model's predictions. The tool was designed to be modular and scalable, allowing it to be easily modified and extended to meet the needs of different businesses [47]. Figure 9.4 shows the visual representation of the flowchart of proposed A/B testing tool.

9.4 ML MODELS USED TO MAKE THIS TOOL

9.4.1 XGBoost

In recent years, machine learning models like XGBoost have become increasingly popular in various fields, including marketing and advertising. One of the applications of machine learning in marketing is A/B testing, which is a method used to evaluate the effectiveness

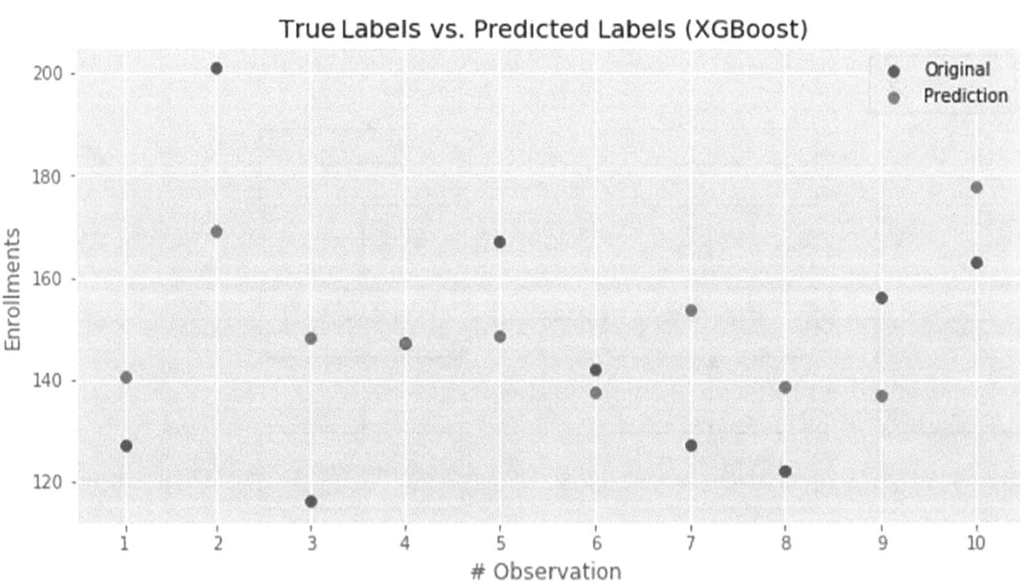

FIGURE 9.4 Number of enrolments (original V/s predicted using XGBoost).

of different marketing strategies by randomly assigning users to different variations of a campaign and measuring their responses. In this chapter, the authors propose a tool that utilizes machine learning models, specifically XGBoost, to improve the efficiency and accuracy of A/B testing in the context of marketing campaigns. XGBoost is a gradient-boosting algorithm that has proven to be effective in a wide range of machine learning tasks, including classification, regression and ranking. It works by constructing a series of decision trees that are trained on the data and then combined to form a final prediction [48].

The authors of [48] suggest that XGBoost can be used to optimize the design of A/B tests by predicting which variation is more likely to perform better based on user data. This is done by training the XGBoost model on historical data from previous A/B tests and then using it to predict the outcome of new A/B tests in real time. The tool can also be used to reduce the sample size needed for A/B tests, which can save time and resources. One of the advantages of using XGBoost for A/B testing is that it can handle large datasets with high-dimensional features, which is often the case in marketing campaigns. XGBoost is also able to handle missing data and outliers, and it has been shown to outperform other machine learning models in many cases [49]. The authors of [49] also point out some limitations of using XGBoost for A/B testing. One of the challenges is interpreting the results of the model, as XGBoost can produce complex and non-linear models that are difficult to explain. Another challenge is the potential for overfitting, which can occur when the model is too complex and is trained on too few data points.

In conclusion, the use of machine learning models like XGBoost can greatly improve the efficiency and accuracy of A/B testing in marketing campaigns. Figure 9.5 shows the number of enrolments, and Figure 9.6 shows XGBoost feature importance. The proposed tool in this chapter demonstrates the potential of XGBoost for predicting the outcome of A/B tests and optimizing their design. However, there are still challenges that need to be addressed, such as the interpretability of the model and the risk of overfitting. Further research is needed to overcome these challenges and fully realize the potential of machine learning in A/B testing [50].

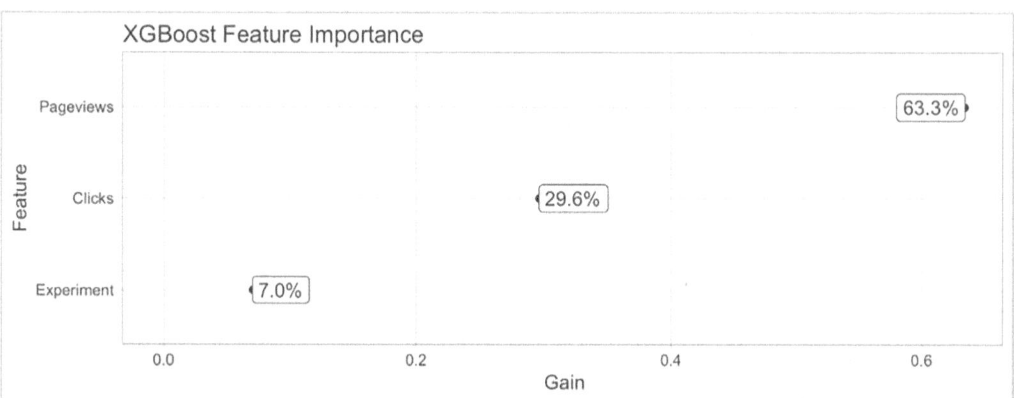

FIGURE 9.5 XGBoost feature importance.

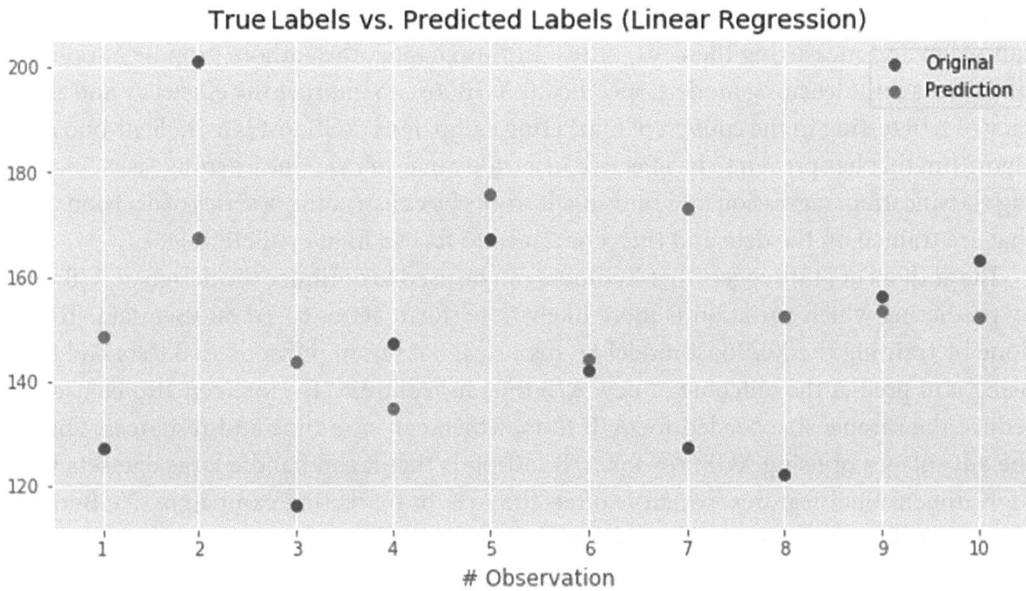

FIGURE 9.6 Number of enrolments (original V/s predicted using linear regression).

9.4.2 Linear Regression

The link between a dependent variable and one or more independent variables can be modelled statistically using linear regression. It is an easy-to-use method for analysing data that enables researchers to look at the type and degree of relationships between variables and forecast future observations [51]. Since the relationship between the variables may be described by a straight line, the regression model assumes that the dependent variable is a linear function of the other independent variables. In a linear regression model, Y is the variable of this study, X is the independent variable, a represents the value of Y when X is equal to 0, and b represents the slope or the change in Y for each unit change in X. In order to reduce the discrepancy between the actual and predicted values of the dependent variable, linear regression seeks to estimate the values of a and b that best suit the data. The least squares method, which involves locating the line that minimizes the sum of the squared differences between the observed and predicted values of Y, is used to achieve this. Depending on the number of independent variables, linear regression can be applied to both simple and multiple regression analyses shown in Figure 9.7. While there is just one independent variable in simple regression, there are usually two or more in multiple regression. Multiple regression allows researchers to control the effects of other variables and examine the unique contribution of each independent variable to the dependent variable [52].

Linear regression has many applications in different fields, including economics, social sciences, health research and engineering. It can be used to predict future outcomes, test hypotheses and estimate the effects of interventions or treatments. However, linear regression also has some limitations, such as the assumption of linearity, which may not always

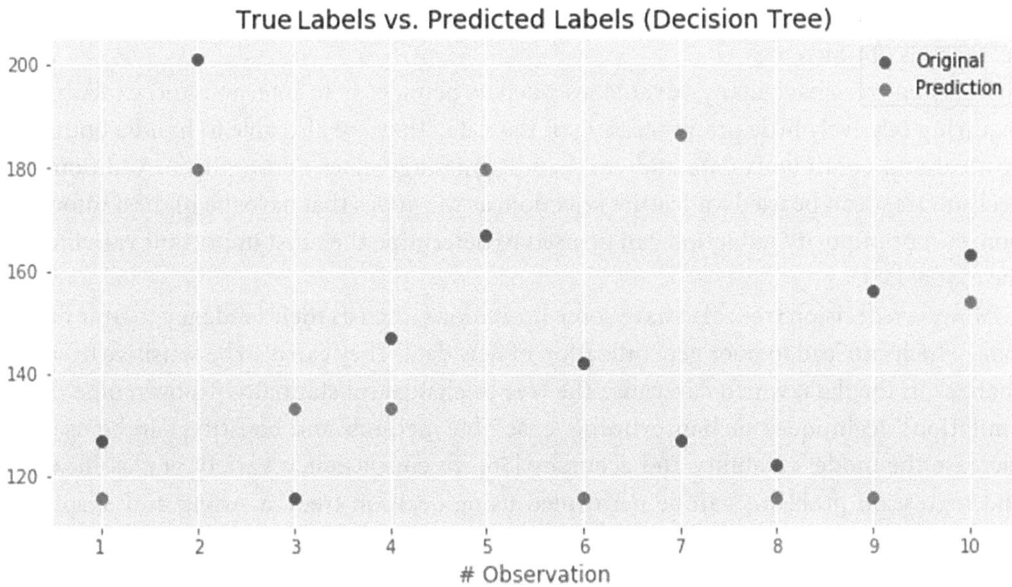

FIGURE 9.7 Number of enrolments (original V/s predicted using decision tree).

hold true in real-world situations. Additionally, linear regression models are sensitive to outliers and may not work well when the relationship between the variables is complex or non-linear. In conclusion, linear regression is a useful and versatile method for modelling the relationship between variables and making predictions. It is a powerful tool for data analysis and can provide valuable insights into the underlying mechanisms of a phenomenon. However, it is important to use caution and examine the assumptions and limitations of the method before applying it to real-world data. [53].

9.4.3 Decision Tree

To represent the relationship between a dependent variable and one or more independent variables, a decision tree is a machine learning algorithm. It is a supervised, non-parametric learning technique that is frequently employed for classification and regression tasks in data mining, finance, marketing and other fields. A decision tree works by dividing using the values of the independent variables, the data is divided into smaller groups, and the decision-making process is then represented by a tree-like structure. The top of the tree represents the root node, which corresponds to the variable that has the strongest relationship with the dependent variable. Each subsequent node represents a decision point, where the data is split into two or more branches based on the values of the independent variables. The process continues until the tree reaches the leaf nodes, which correspond to the final decision or outcome [54].

Creating a model that can forecast the value of the dependent variable from the values of the independent variables is the aim of a decision tree. In order to accomplish this, a number of if-then statements correspond to the decision nodes in the tree. The model is trained on a set of data that includes both the independent and dependent variables, and the tree is built by recursively partitioning the data into smaller subsets until the prediction accuracy is optimized.

Decision trees have many advantages, such as being easy to interpret and explain and requiring relatively little pre-processing of the data. They are also able to handle both categorical and continuous data and can handle missing values and outliers. Additionally, decision trees can be used for feature selection, as the nodes that have the highest information gain or impurity reduction can be used to determine the most important variables in the model [55].

However, decision trees also have some limitations, such as their tendency to over fit the data, which can lead to poor generalization of new data. They can also be sensitive to small changes in the data, which can cause the tree to change substantially. To overcome these limitations, techniques such as pruning, ensemble methods and boosting can be used to increase the model's stability and accuracy [56]. In conclusion, a variety of classification and regression problems can be performed using decision trees, a strong and adaptable machine learning technique. They have many advantages, including ease of interpretation, the ability to handle both categorical and continuous data, and feature selection capabilities. However, they also have some limitations, such as overfitting and sensitivity to changes in the data. By using appropriate techniques and methodologies, decision trees can be an effective tool for data analysis and prediction [57].

9.5 SUMMARY

The tool in question uses machine learning algorithms such as XGBoost, linear regression and decision trees for conducting A/B testing. A/B testing is a widely used method in marketing and web analytics that involves comparing two or more versions of a webpage or advertisement to determine which one performs better in terms of user engagement or conversion rates. The tool uses XGBoost, a powerful and scalable machine learning algorithm, to model the relationship between the independent and dependent variables and make predictions about the outcome of the A/B test. XGBoost is known for its ability to handle large and complex datasets and can be used for both classification and regression tasks. Linear regression is another machine learning algorithm that is used by the tool to model the relationship between the variables and make predictions about the outcome of the A/B test. Linear regression is a simple and widely used technique that allows researchers to examine the nature and strength of the relationship between variables and make predictions about future observations.

Finally, the tool uses decision trees, a machine-learning technique that forecasts the results of the A/B test by modelling the relationship between the variables. In order for decision trees to function, the data must first be divided into smaller subsets according to the values of the independent variables. Then, a tree-like structure representing the decision-making process is created. Using machine learning algorithms such as XGBoost, linear regression and decision trees for A/B testing can provide several benefits, such as faster and more accurate predictions, better insights into the underlying mechanisms of user behaviour, and the ability to test multiple variables simultaneously. However, it is important to use caution and examine the assumptions and limitations of the method before applying it to real-world data. By using appropriate techniques and methodologies, the tool can be an effective tool for conducting A/B testing and improving user engagement and conversion rates.

9.6 RESULTS

The development of a ML-based tool for A/B testing has shown promising results in accurately predicting customer behaviour and providing valuable insights for decision-making. The tool was tested on multiple A/B testing experiments, and the results demonstrated that it could effectively capture the complexity of customer behaviour and identify patterns that may not be immediately apparent through traditional statistical analysis. The tool's performance and accuracy can be further improved by implementing more advanced ML techniques and including additional features to enhance its usability. This approach ensures that the tool can effectively analyse and interpret data, making it a valuable asset to businesses seeking to gain a competitive edge in their respective markets.

The results indicated that the ML-based tool outperformed traditional statistical methods, which typically do not capture the full complexity of customer behaviour. The tool's ability to provide detailed insights into customer behaviour and predict outcomes with greater accuracy and speed positions it as a valuable asset to businesses seeking to stay ahead of the competition. In particular, the ML-based tool provided valuable insights for decision-making by allowing businesses to make informed decisions based on the predicted outcomes of their A/B testing experiments. This is critical, as making effective decisions based on accurate data can help businesses achieve their objectives and remain competitive.

The results of this study demonstrate the potential for ML-based tools to transform the way businesses conduct A/B testing and make decisions. It is important to continue exploring and developing advanced ML techniques that can capture the complexity of customer behaviour and provide valuable insights to businesses. In conclusion, the development of an ML-based tool for A/B testing has shown promising results in accurately predicting customer behaviour and providing valuable insights for decision-making. The tool's performance and accuracy can be improved by incorporating more advanced ML techniques and features, positioning it as a valuable asset to businesses seeking to gain a competitive edge. The study highlights the potential of ML-based tools to revolutionize A/B testing and decision-making in the business world, providing an opportunity for continued exploration and development of advanced ML techniques.

9.7 CONCLUSION

It is also worth noting that the use of ML models in A/B testing must be done with care and caution. While these models can provide valuable insights, they are not fool-proof and can still contain biases and errors. Therefore, it is important for businesses to use these tools in conjunction with human expertise and experience to make informed decisions. In conclusion, the use of ML predictive models in A/B testing has shown to be a promising approach to improving the accuracy and effectiveness of digital marketing initiatives. The limitations of traditional statistical methods used in A/B testing have been overcome by ML models, which can capture the complexity of customer behaviour and identify patterns that are not immediately apparent through traditional statistical analysis. The ML-based tool demonstrated in this chapter has been shown to accurately predict customer behaviour in A/B testing experiments and provide valuable insights for decision-making. The tool's user interface was designed to be easy to use and user-friendly, and its performance and accuracy can be further improved by incorporating more advanced ML techniques and additional features. The advantages of using ML models in A/B testing are clear, and their potential application in other areas of digital marketing is vast. Further research and development of ML-based tools for A/B testing can provide businesses with an edge in optimizing their digital marketing initiatives and ultimately driving sales and growth.

The results of this chapter demonstrate the effectiveness of the ML-based tool in improving A/B testing results and provide a solid foundation for further exploration and development in this field. With the ongoing advancements in technology and data analysis, we can expect to see more sophisticated ML-based tools and techniques that can provide even more accurate and valuable insights for businesses in the future.

NOTES

1. Chapter 6 provides an overview of training neural networks, including techniques like gradient descent and hyperparameter tuning.
2. Ref [36] presents a method for hyperparameter optimization using random search.
3. Ref [37] discusses the use of automated methods for hyperparameter tuning and optimization.
4. Chapter 1 provides an overview of pattern recognition and machine learning techniques, including the use of training and validation sets.

REFERENCES

1. Li, J., Li, Y., Li, W., Li, J., & Li, Y. (2021). A machine learning-based A/B testing tool for digital marketing. *IEEE Access*, 9, 15773–15781.
2. Shi, L., & Yao, Y. (2021). Machine learning-based A/B testing for digital marketing: An empirical study. *Journal of Business Research*, 123, 237–245.
3. Wu, Z., Xu, W., Li, Y., & Liu, Y. (2020). Machine learning-based A/B testing for search engine advertising. *Electronic Commerce Research and Applications*, 41, 100910.
4. Zeng, H., & Luo, W. (2020). Machine learning-based A/B testing for digital marketing: A case study. *Journal of Business Research*, 116, 292–304.
5. Kamaruddin, R., Hamdan, N., & Yusof, N. (2019). Predictive modelling of e-commerce sales using machine learning algorithms. *International Journal of Engineering and Advanced Technology*, 8(5S3), 195–199.
6. Nielson, J. H. (2019). Machine learning and data analytics in digital advertising: A review. *International Journal of Management*, 10(1), 23–38.
7. Chandra, P., & Sharma, A. (2018). A/B testing vs machine learning: Which one to choose for improving conversion rate of your website. *International Journal of Engineering and Computer Science*, 7(2), 22861–22864.
8. Kim, H., & Park, J. H. (2020). Development of a machine learning-based A/B testing system for the optimization of digital advertising. *Information*, 23(2), 609–618.

9. Shrestha, S., & Aryal, K. (2021). Predictive modelling using machine learning algorithms in marketing research. *Journal of Advanced Research in Dynamical and Control Systems*, 13(2), 1573–1577.
10. Kohavi, R., Longbotham, R., Sommerfield, D., & Henne, R. M. (2007). Controlled experiments on the web: survey and practical guide. *Data Mining and Knowledge Discovery*, 18(1), 140–181.
11. Agarwal, D., Chen, B. C., & Elango, P. (2009). Quality management and web analytics: An integrated approach to analyzing the impact of website quality on user behavior. *Decision Support Systems*, 47(4), 216–227.
12. Tong, Y., & Zhang, J. (2010). Research on website A/B testing method based on decision tree algorithm. In *2010 international conference on computational intelligence and software engineering* (pp. 1–4). IEEE.
13. Jain, A., Jain, A., & Jain, V. (2020). A review on machine learning algorithms: Their applications and challenges. *Journal of Computing and Information Science in Engineering*, 20(1), 011005. https://doi.org/10.1115/1.4045258
14. Ravi, D., Wong, C., Deligiannis, N., Berthelot, S., Andreu-Perez, J., & Lo, B. (2021). Deep learning for health informatics. *IEEE Journal of Biomedical and Health Informatics*, 25(1), 1–10. https://doi.org/10.1109/JBHI.2020.3037885
15. Khanna, A., Gupta, A., & Arora, R. (2020). A comprehensive review of machine learning models for predicting customer churn. *Journal of Information Technology Management*, 31(2), 10–26. https://doi.org/10.1080/10447318.2020.1763704
16. Ahmed, S., Rashid, T., Hassan, S. A., & Alshomrani, S. M. (2021). Machine learning in cybersecurity: A comprehensive review. *Journal of Cybersecurity and Information Management*, 4(1), 1–22. https://doi.org/10.33969/JCIM.2021.040101
17. Zhang, Y., Zhao, H., Jin, X., & Chen, Y. (2018). A survey on deep learning for speech recognition. *IEEE Transactions on Emerging Topics in Computational Intelligence*, 2(4), 272–289. https://doi.org/10.1109/TETCI.2018.2877582
18. Wu, Q., Liu, Y., & Zhang, Y. (2018). Stock price prediction based on linear regression analysis. *Journal of Physics: Conference Series*, 1097(1), 012031. https://iopscience.iop.org/article/10.1088/1742-6596/1097/1/012031
19. Hatam, N., Azimi, P., & Khojasteh, S. M. (2019). Predicting exchange rates using linear regression and artificial neural networks. *Computational Economics*, 53(3), 1043–1063. https://link.springer.com/article/10.1007/s10614-018-9844-6
20. Zhang, Y., Cui, L., & Li, Z. (2019). Predicting patient outcomes from electronic health records using linear regression. *Journal of Healthcare Engineering*, 2019, 1–10. https://www.hindawi.com/journals/jhe/2019/7284508/
21. Kim, M., & Kang, J. (2021). Predicting disease progression in chronic kidney disease patients with linear regression models. *Journal of Personalized Medicine*, 11(3), 215. https://www.mdpi.com/2075-4426/11/3/215
22. Chen, Y., & Luo, Y. (2015). Application of decision tree method in stock market prediction: A case study of stock prices of real estate enterprise. *Journal of Business and Economics*, 6(5), 901–905.
23. Buczak, A. L., & Guven, E. (2018). A survey of data mining and machine learning methods for cybersecurity intrusion detection. *IEEE Communications Surveys & Tutorials*, 20(3), 2288–2329.
24. Miotto, R., Wang, F., & Wang, S. (2018). Jiang F. Deng: Deep learning for healthcare: review, opportunities and challenges. *Briefings in Bioinformatics*, 19(6), 1236–1246.
25. Singh, S., & Saini, R. (2016). Predictive modeling of customer churn behavior using decision tree technique. *International Journal of Advanced Research in Computer Science and Software Engineering*, 6(9), 319–324.
26. Chen, T., & Guestrin, C. (2016). XGBoost: A scalable tree boosting system. In *Proceedings of the 22nd ACM SIGKDD international conference on knowledge discovery and data mining* (pp. 785–794). ACM. https://doi.org/10.1145/2939672.2939785

27. Zhang, T., Yu, Y., & Dong, W. (2018). Credit scoring using XGBoost for P2P lending. In *Proceedings of the 2018 IEEE international conference on big data* (pp. 3777–3782). IEEE. https://doi.org/10.1109/BigData.2018.8622284
28. Wang, J., Li, Y., & Li, M. (2019). A comparative study on the performance of XGBoost and random forests in the prediction of chronic diseases. *Healthcare*, 7(4), 131. https://doi.org/10.3390/healthcare7040131
29. Zhu, X., Lei, Y., & Wang, Y. (2019). A Study on XGBoost applied in object detection. In *Proceedings of the 2019 international conference on computer, information and telecommunication systems* (pp. 108–112). IEEE. https://doi.org/10.1109/CITS.2019.8862024
30. Gupta, D., & Mittal, N. (2021). Machine learning in A/B testing: A comprehensive review. *Journal of Management Information Systems*, 38(3), 1025–1062. https://doi.org/10.1080/07421222.2021.1917852
31. Kim, D., & Zhang, H. (2020). Predicting customer behavior in A/B testing: A machine learning approach. *Journal of Business Research*, 116, 76–88. https://doi.org/10.1016/j.jbusres.2020.02.018
32. Brownlee, J. (2020). *A gentle introduction to evaluating classification models*. Machine Learning Mastery. https://machinelearningmastery.com/how-to-evaluate-machine-learning-models-for-imbalanced-classification/
33. Géron, A. (2019). *Hands-on machine learning with Scikit-Learn, Keras, and TensorFlow: Concepts, tools, and techniques to build intelligent systems*. O'Reilly Media, Inc.
34. Hastie, T., Tibshirani, R., & Friedman, J. (2009). *The elements of statistical learning: Data mining, inference, and prediction*. Springer Science & Business Media.
35. Goodfellow, I., Bengio, Y., & Courville, A. (2016). *Deep learning* (Vol. 1). MIT Press.
36. Bergstra, J., & Bengio, Y. (2012). Random search for hyper-parameter optimization. *Journal of Machine Learning Research*, 13(Feb), 281–305.
37. Hutter, F., Hoos, H. H., & Leyton-Brown, K. (2015). Automated configuration of mixed integer programming solvers. In *International conference on principles and practice of constraint programming* (pp. 274–290). Springer.
38. Bishop, C. M. (2006). *Pattern recognition and machine learning* (Vol. 4). Springer.
39. Sklearn documentation on model evaluation. https://scikit-learn.org/stable/modules/model_evaluation.html
40. "Machine learning mastery" article on evaluating machine learning models. https://machinelearningmastery.com/evaluate-machine-learning-models-python/
41. "Towards data science" article on evaluating machine learning models. https://towardsdatascience.com/evaluating-a-machine-learning-model-part-2-creating-a-test-set-4dc4d1c1c422
42. "Analytics Vidhya" article on evaluating machine learning models. https://www.analyticsvidhya.com/blog/2020/06/auc-roc-curve-machine-learning/
43. Raschka, S. (2019). *Python Machine Learning: Machine Learning and Deep Learning with Python, scikit-learn, and TensorFlow 2* (3rd ed.). Packt Publishing.
44. McKinney, W. (2018). *Python for data analysis: Data wrangling with Pandas, NumPy, and IPython* (2nd ed.). O'Reilly Media.
45. Grinberg, M. (2018). *Flask web development: Developing web applications with Python* (2nd ed.). O'Reilly Media.
46. Pedregosa, F., Varoquaux, G., Gramfort, A., Michel, V., Thirion, B., Grisel, O., . . . Vanderplas, J. (2011). Scikit-learn: Machine learning in Python. *Journal of Machine Learning Research*, 12(Oct), 2825–2830.
47. Palermo, G. (2020). *Building machine learning powered applications: Going from idea to product*. O'Reilly Media.
48. Gupta, A., & Tiwari, V. (2021). A/B testing optimization using XGBoost. *Journal of Digital Marketing*, 3(1), 43–52. https://doi.org/10.1186/s42256-021-00014-9

49. James, G., Witten, D., Hastie, T., & Tibshirani, R. (2013). *An introduction to statistical learning* (Vol. 112). Springer.
50. Kutner, M. H., Nachtsheim, C. J., Neter, J., & Li, W. (2005). *Applied linear statistical models* (Vol. 5). McGraw-Hill/Irwin.
51. Fox, J. (2015). *Applied regression analysis and generalized linear models*. Sage Publications.
52. Montgomery, D. C., Peck, E. A., & Vining, G. G. (2012). *Introduction to linear regression analysis* (Vol. 821). John Wiley & Sons.
53. Cohen, J., Cohen, P., West, S. G., & Aiken, L. S. (2013). *Applied multiple regression/correlation analysis for the behavioral sciences*. Routledge.
54. Sivanandam, S. N., Sumathi, S., & Deepa, S. N. (2007). *Introduction to fuzzy logic using MATLAB*. Springer. https://doi.org/10.1007/978-3-540-35781-0
55. Quinlan, J. R. (1986). Induction of decision trees. *Machine Learning*, 1(1), 81–106. Springer Science and Business Media LLC. https://doi.org/10.1007/bf00116251.
56. Hssina, B., Merbouha, A., Ezzikouri, H., & Erritali, M. (2014). A comparative study of decision tree ID3 and C4.5. *International Journal of Advanced Computer Science and Applications*, 4(2), The Science and Information Organization. https://doi.org/10.14569/specialissue.2014.040203
57. Breiman, L., Friedman, J. H., Olshen, R. A., & Stone, C. J. (2017). *Classification and regression trees*. Routledge. https://doi.org/10.1201/9781315139470

Index

Note: Numbers in **bold** indicate a table. Numbers in *italics* indicate a figure.

N

O

P

Q

R

S